THE SLEEPYHEAD'S BEDSIDE COMPANION

Sean Coughlan is a journalist at the BBC. Since childhood he has had a deep dislike of getting up too early. He lives in London and as the father of three young children, he takes a very keen interest in getting enough sleep.

Sean Coughlan

The Sleepyhead's Bedside Companion

arrow books

This paperback edition published by Arrow Books 2010

10 9 8 7 6 5 4 3 2 1

First published in Great Britain in 2009 by Preface Publishing

Arrow Books
20 Vauxhall Bridge Road
London SW1V 2SA

An imprint of The Random House Group Limited

www.rbooks.co.uk

Addresses for companies within The Random House Group Limited can be found at www.randomhouse.co.uk

The Random House Group Limited Reg. No. 954009

A CIP catalogue record for this book is available from the British Library

ISBN 978 1 84809 174 0

Mixed Sources
Product group from well-managed forests and other controlled sources
www.fsc.org Cert no. TT-COC-2139
© 1996 Forest Stewardship Council

The Random House Group Limited supports The Forest Stewardship Council (FSC), the leading international forest certification organisation. All our titles that are printed on Greenpeace-approved FSC-certified paper carry the FSC logo. Our paper procurement policy can be found at www.rbooks.co.uk/environment

Printed and bound in Great Britain by CPI Bookmarque, Croydon CR0 4TD

To my mother and father – in thanks for inventing the
Eight O'Clock Man to make me sleep at night

Come, Let's to Bed

To bed, to bed
Says Sleepyhead,
Tarry a while, says Slow.
Put on the pan,
Says Greedy Nan,
We'll sup before we go.

CONTENTS

INTRODUCTION

In Praise of Sleep

SLEEP HAS BECOME A MODERN OBSESSION. In our overcrowded, time-starved lives, it's sleep that we crave more than almost anything else. When we run out of time, it's sleep that suffers, turning tiredness and sleep deprivation into a daily self-inflicted punishment.

What makes this an even greater folly is that there is so much to be enjoyed about sleep. It's nature's finest and most mysterious free gift. It's an inexhaustible source of rest and recuperation, a respite from the hassle and hustle all around. Instead of rationing sleep, we should be revelling in it.

In a burnt-out 24-hour culture, sleep is a last patch of long, cool, green grass, a place to catch your breath and look up at the sky. Sleep is an essential part of the natural habitat of being human. It feels like the only place left where no one expects you to work or shop.

So why do we treat sleep so badly? Why don't we savour sleep, enjoy its pleasures, learn about its history, culture and meaning? If sleep was charged at restaurant prices then maybe we'd have sleep gourmets and sleep recipes, turning the nightly kip into an epicurean event. Instead we chop back sleep, cutting its corners, doing without.

It's not as if we don't know the consequences. The dangers of not getting enough good quality sleep have become ever more apparent. Barely a month passes without some new medical research showing the physical damage caused by sleeplessness. Lack of sleep makes us anxious, irritable and unable to concentrate. Only getting five hours' sleep leaves someone as impaired in performing simple tasks as if they were over the drink-driving limit. Over a longer period, inadequate sleep sharply increases the risk of heart disease and is associated with an increased likelihood of obesity and diabetes. Sleep deprivation makes it more difficult to learn and remember.

So why when we know all this do we make a mess of sleep? How do we turn it into such a disaster area, with so many claiming to be sleep starved or suffering from insomnia or the growing problem of sleep apnea? Everyone sleeps, maybe not as much as we might like, but it's a universal human experience, as instinctive as breathing. But for something that should come so naturally, we seem to have turned it into a problem.

Sleep has been caught in a pincer movement. There are so many demands and distractions in our restless, broadband lives that we go to bed too late and get up too early and never get the afternoon sleep that our bodies want as compensation. If that isn't bad enough, then there is a misguided character assassination of sleep, presenting it as a waste of time, an enemy of hard work and ambition. The fact that sufficient sleep is a physical necessity and that we would drop dead sooner from lack of sleep than lack of food doesn't seem to register.

It's not as if we have much of an excuse to miss out on sleep. They might have privatised the rainwater and polluted the fresh air, but sleep is there in abundance and without cost. No charge for the re-charge. Sleep is the spa that never closes. It's completely egalitarian. Samuel Johnson called it the 'impartial benefactor', ready to come to the rescue of saint and sinner, prince or prisoner alike. Sleep has no priority boarding queue. But still we manage to get it wrong, cutting short our sleep and having a miserable day of suffering from what doctors describe as 'TATT' – or 'tired all the time'.

We've all been there when sleep means more than anything else. Sitting exhausted in some miserable departure lounge when a flight has been cancelled, the lack of sleep leaving you feeling rougher than the carpet tiles on the airport floor. Or there's the exhaustion of trying to calm a howling baby on a winter's night, with the red eyes of daylight appearing over the rooftops. Or that dead-head sensation of going to work with so little sleep that it almost hurts. All we want at such times is for sleep to come to our rescue, to help the weary and worn-out, to throw its cloak around us.

This book is about restoring the balance, celebrating the neglected glory of a good night's sleep. Instead of worrying about sleep, or dismissing it as wasted hours, we should be treating it as an intriguing delicacy, a pleasure that we can enjoy again and again, a mystery train that takes us into dreams and adventures, giving us a tour around our own subconscious, a place with its own history and forgotten culture. We should admit that there are few finer pleasures than that moment when we give in to the desire to sleep.

Sleep connects us with something very profoundly rooted in nature, it's where we let go, sharing the rhythm of sleeping and waking with all kinds of other living creatures, stepping away from the shrill demands of the day. It's not worth getting into bed for anything less.

This is a bedtime story for the sleepyhead inside all of us.

Marie Antoinette rested her head in ostentatious splendour . . . before losing it on the guillotine.

And So To Bed

Undercover: a history of the bed

WHERE IS THERE a finer invention than the bed? It's a machine for sleeping. It needs little maintenance and delivers magnificent results. It is the ultimate four-legged friend. But do we celebrate its history, is there a statue in every town to its pioneers?

Wherever there have been people there have been beds. In the Neolithic era, people slept under skins and furs in beds lined with grass or heather. On the rocky Orkneys, in the 5,000-year-old archaeological site at Skara Brae, there are ancient box-shaped beds made out of stone, which would have been softened with animal pelts and bracken.

A history of an object usually explains how much its design and performance have been transformed over the centuries. But the curious splendour of the bed is contained in how little it has changed. It could have been all kinds of shapes, but look inside an ancient Egyptian tomb and what does the bed look like? It's a wooden rectangular frame with four short legs. You could buy something similar in John Lewis. Of course the pharaohs had an appetite for decoration, so there are more elaborate beds with all kinds of animal designs attached, but the basic template has remained unchanged.

The Roman bedroom was known as the cubiculum, and this was often rather cubicle-like and functional. The type of beds the Romans brought on their conquest of Britain were also plain and practical, not dissimilar to the beds brought by later Scandinavian invaders, known as the tribe of Ikea. The Roman bed was a rectangular frame, made out of wood or metal, with straps or ropes or metal bands fixed across the top to support some kind of soft bedding stuffed with feathers or straw.

The word 'bed', as direct and simple as the thing it describes, was introduced by the Anglo-Saxons. There are references to bed-curtains in Saxon writing, suggesting that these would have been hung around the bed to keep in the warmth. Pillows were stuffed with straw, and bed-covers could be made from goatskin or bearskin. The bed would have been an important possession for a powerful individual. A recent excavation of a seventh-century burial site in Yorkshire found a woman's body draped in jewellery and lying in a bed. This burial bed, made for the sleep of death, was constructed from wood and held together with iron.

Edward the Confessor's deathbed is stitched into the story told in the Bayeux Tapestry, which shows him resting on a bolster-type pillow, his wooden bed decorated with carvings in the shape of animal heads. These are similar to Viking-style beds found in Norway, with carvings reminiscent of the figureheads on the prow of a longboat.

The Normans, more into chainmail than comfort, continued in the tradition of plain bedding. But like tourists suddenly dazzled by something exotic, they were responsible for one of the great leaps forward in bed making. The Crusaders, used to their draughty bedrooms in unheated castles, came across very different sleeping arrangements in the Middle East. There people were sleeping in tents and sun-filled houses on soft cushions, surrounded by silks and sensuous materials. There was even a new word to describe this relaxing style, an Arabic word for 'the place where something is thrown', describing the way people might sleep on comfortable cushions thrown on the floor. This strange new word was 'mattress'.

Inspired by their overseas adventures, the wealthy and powerful began to assemble more elaborate sleeping places. Such beds had a canopy above them, called a 'tester', and sides covered with expensive cloths and textiles. The bed became a status symbol, often the most valuable bit of furniture that a family could own.

zzzzzzzzzzzzzzzz

The bed became a status symbol, often the most valuable bit of furniture that a family could own.

In the fourteenth century, the four-poster bed appeared, a piece of furniture that was a solid statement of power and wealth. These beds were made even more comfortable by feather-filled mattresses, which were imported from the stylish French. When medieval monarchs were on the road, they brought these mighty beds with them, dismantling and then erecting them wherever they were going to stay. The wooden frame would be surrounded by embroidered cloths, studded with jewels and decorated with textiles brought from the limits of the known world. The king and queen would lie enclosed in the middle of all this splendour, like little pearls in a richly ornamented shell.

But when the monarch suffered a sudden loss of popularity, or even the sudden loss of a head, the bed became one of the first targets for looters and free-loaders, who would strip it back to the bare boards. In the *Gentleman's Magazine*, a mid-Victorian publication, there is a marvellously embroidered account of what happened to Richard III's bed after he had lost the Battle of Bosworth in 1485. The defeated king's bed was set upon by robbers and soldiers, who tore off everything valuable they could carry, rushing away with precious stones and luxurious materials. But the wooden frame was too big and heavy to remove and ended up in the possession of a Leicester innkeeper, who, pub-landlord style, used it as his own rather flashy bed.

The bed stayed in this pub for the following century, passing from tenant to tenant, until one of the landlords discovered why the bed was so heavy: it had a secret compartment stuffed with the king's gold coins. This windfall of money soon saw a rapid improvement in his social standing, with this 'man of low condition' being made the local chief magistrate. In a similarly predictable fashion, everyone fell out over the money when he died. His widow was murdered, the coins were stolen and seven people were hanged and one burned to death for the conspiracy. Such was the unhappy legacy of the 'last abiding-place of the last monarch of the Middle Ages'.

The account contains the following description of how the bed looked:

> 'Richly and curiously carved in oak, with fleur-de-lys profusely scattered over it, its panels inlaid with black, brown and white woods, the styles consisting of Saracenic figures in high relief, it proves from the singularity of its construction, the true purpose for which it was designed, every portion of it but the body being fabricated to take to pieces and put up at will; so that for travelling it speedily became transformed into a huge chest.'

The Victorian gents reading this story in their leather armchairs certainly had some imagination-stirring details to savour. Richard had woken in opulent surroundings, but by the end of the day his bed was being torn apart by peasants and his dead body was lying in a church 'as naked as ever he was born'. The fact that the legend might not be entirely reliable shouldn't spoil the story.

These richly decorated, heavily canopied beds were pockets of warmth in unheated houses. In paintings, where beds regularly appeared as props for idling goddesses or in poignant deathbed scenes, you can see them with heavy bedspreads, long curtains and brocaded canopies. The cold air was repelled and a cocoon of warmth created within. In warmer countries such as Italy, beds glimpsed in the backgrounds of paintings are shown without such curtains. In the 1320s, a Lorenzetti religious painting shows a bed, bedspread, sheets and pillow that could be in any modern house; a century later, a painting by Sano di Pietro shows something similar.

Meanwhile, sleeping quarters for the poor could mean curling up in a corner on straw or under a few blankets or wherever they could stay warm. The rural poor, trying to earn a living in the harvest, might be sleeping in barns or tucked into hayricks. Accounts written at the time tell of these poor itinerant workers lying like scarecrows in the straw.

But the middle classes were getting excited about beds. The Habitat gene was beginning to show itself. The Elizabethan urban middle class wanted bigger beds. The stout yeomen had had plain no-nonsense wooden beds for centuries, but now there was an appetite for something more elaborate. A comfortable merchants' house was likely to contain several tester beds with embroidered drapes. Mattresses would have been stuffed with feathers or wool.

There was also a fashion for ridiculously large beds. The Great Bed of Ware, probably built for an inn at Ware, Hertfordshire, around 1590, was big enough for a dozen people. It became a tourist attraction and was mentioned in *Twelfth Night.* Not to be outshone by her subjects, Queen Elizabeth ordered a bed that was to be as lavish as it was enormous, ornamented with gold, silver and ostrich feathers. However, this grandiose construction proved to be of little consolation to Elizabeth in her last days. Contemporary accounts of her death recorded that she refused to lie down in her bed and instead chose to rest on the floor, supported by a few cushions.

No such anxieties affected the French monarch Louis XIV whose state bed at Versailles showed how the Sun King slept in golden splendour. The state bed was a centrepoint of royal ceremony, where the king might receive guests. There were five separate categories of bed, divided according to the degree to which they were for public show and for private use. It is claimed that Louis XIV had about four hundred beds altogether.

There were still plenty of people not enjoying such comfort. Servants might have got off the straw, but they were often confined to modest 'truckle beds' which could be pulled out for the night.

There were other practical considerations. An advert from a bedmaker in the Strand in London in the 1790s announced a 'new and improved style of Four-post and Tent beds' which promised to 'really prevent the harbour of Vermin'. The same rat-repelling advert offered 'improved sofa beds', which sound surprisingly modern.

By the middle of the nineteenth century there were other considerations. In more modest Victorian houses these big beds with their canopies and curtains took up a great deal of space. There were also worries about hygiene and the amount of dust gathered by these big, baggy, mahogany monsters. A healthier, fashionable alternative arrived in the form of mass-produced brass beds, without any side coverings and sold with sprung mattresses.

But that didn't mean that everyone had their own bed. Victorian social reformers campaigning against over-crowding found examples of lodging houses in Lancashire in the 1850s where sixteen people were sharing a single bedroom. In the 1860s, two-bedroom houses rented to working-class families in industrial cities often had ten or more occupants.

zzzzzzzzzzzzzzzzzz

There were five separate categories of bed, divided according to the degree to which they were for public show and for private use. It is claimed that Louis XIV had about four hundred beds altogether.

The Victorians used the word 'chumming' to describe the practice of the overcrowded sharing of a bed or living space, such as prisoners who might be 'chummed' in a jail cell. It was a word that started out as a university term for sharing a room and seemed to have gone downmarket. A more pungent older word for sharing a bed with several people was 'pigging'.

Sharing bedrooms and beds was part of everyday life. This could be a malodorous experience, according to accounts of taverns in eighteenth-century America, where it was considered usual practice for guests to share a bed with strangers. There are complaints about having to share a bed with the 'greasy landlord' or with a man who made 'hideous noises' below the bedclothes. A family of nine was able to share one bed, 'all pigged together lovingly'.

The fact that 'sleeping together' now means such a different thing is indicative of how much our sleeping habits have changed. The twentieth century saw the privatisation of sleep, with a more affluent society expecting that individuals and couples should have their own beds. The bed itself became less of a status symbol, as people looked to cars and gadgets and house improvements as more visible ways of flaunting their success.

There were some further technological changes in the modern bed – memory foam and nylon friction burns were just two innovations. But really the story of the bed is one of glorious continuity. Go to a museum in Rome and you can trip over an Etruscan bed frame, then go a couple of blocks into a furniture shop and there will be something not that different on sale.

The real marvel is that such a simple device has been the cradle of so much of life. The French writer Isaac de Benserade described it as the theatre of laughter and tears. Samuel Johnson provided his own translation: 'In bed we laugh, in bed we cry; And born in bed, in bed we die.'

Let's salute the neglected bedmakers and mattress stuffers. They've worked hard for centuries so that we can rest in peace.

See also Hot in bed: the electric blanket, p. 70

Gourmet sleep recipes

THE APPETITE FOR SLEEP is rich and varied. Here is an entirely unscientific collection of five classic sleeps.

Christmas afternoon

You've eaten a Christmas dinner so vast that your entire body is required for the digestive effort. Almost all your mental capacity is now devoted to breaking down Alpine slabs of Christmas pudding. Everyone has that glassy look that overcomes old people who have been drinking at weddings. With some careful navigation, the armchair is reached, and you lower yourself gently into its supportive arms. It takes too much effort to lift the remote control and through hooded eyes you watch the opening credits of *The Snowman*. He's walking in the air. Again. No adult has ever seen the middle section of *The Snowman*, because after a lively start – flying snowmen, interesting aerial views of Brighton – the lullaby chords create an irresistible urge to sleep. Give in to this urge. Your body is in a digestive paralysis and your stomach looks like a photo of a snake that's eaten a sheep. Your senses are being lulled. Every ounce of your body is calling out for sleep. Give in to this most delicious moment. This is going to be the best Christmas present you will get.

Parents' sleep paradise

If you ever go to a conference or work event that involves a night away, watch out for the parents of young children.

While everyone else is hiding their wedding rings and looking forward to a night on the drink, the away-day parent has got an entirely different ambition. They've been thinking about that hotel bed for months, imagining its firm contours, the pert corners of a plump pillow, the snap of fresh bedclothes smoothed into a welcoming neatness. There might even be a chocolate on the covers and some rubbish about saving the environment by not using too many bath towels. That big beautiful bed has got their name written on it. For months and months they've had their sleep disturbed by a child, this lack of sleep has become an obsession, there is nothing they crave more than a full night of uninterrupted sleep. So while everyone else is heading for the bar, they make an excuse about checking some e-mails and say they'll catch the partygoers later. No chance. At an indecently early hour, they shake off everyone else and head straight up for the bedroom, fantasising about the luxury of being asleep with nothing to wake them. This is the sweetest moment, the fulfilment of a pleasure that has been denied for so long, the long chilled glass of water in the middle of a desert. Do not disturb.

First morning of the holiday

If Schadenfreude were a lovely little holiday village just outside Salzburg, I think I might like to pay a visit there. Because waking up on the first morning of a holiday is a particularly selfish pleasure. A quick check of the watch and you can guess what all the other poor suckers at work are doing at that very moment. For you, my friend, the war is over. The holiday is here, there is no work, this pillow is waiting for a quality relationship with the side of your face.

This lie-in could last for ever, the whole holiday stretches out ahead of you. This is a sweet moment of sleep nirvana.

Lazing on a sunny afternoon

There is a special feral pleasure in falling asleep outside. Waking up below the sky and clouds feels different from coming to and looking up at the ceiling. The feel of the ground and the smell of grass add to this suburban exotica, roughing it a few dozen feet away from the back door. It might not exactly be disappearing into the long grass of the steppes, but the back garden or any patch of the outdoors can deliver a succulent slice of sleep. The breeze on your face feels fresh, it's less stuffy than indoors and feels more natural. As a species, we must have spent many thousands of years sleeping outdoors before the double bed was invented. The sounds and scents are there to add to the lullaby. The soothing rattle of trains, tennis commentary from next door's television, insects dive-bombing flowers, the screams and splashing of paddling pools, the scent of suntan lotion and barbecues. My eyes are beginning to feel heavy even thinking about it.

z z z z z z z z z z z z z z z z z z z

This is the sweetest moment, the fulfilment of a pleasure that has been denied for so long, the long chilled glass of water in the middle of a desert. Do not disturb.

Asleep before your head hits the pillow

Of course, the sweetest dishes are those you can never taste again, those that stay in the memory. No one sleeps better than they did in childhood, after long days spent running around outside, evenings that seemed to go on for ever, your face hot from the sun, on a summer holiday that never seemed to end. It might be nostalgia, it might be romanticising a golden era that never existed, but there is something special about the completeness of a child's sleep. They run around for hours and hours and then, as though flicking a switch, they fall into deepest sleep. When my young daughters fall asleep like this they are completely oblivious to the world around them. They can be carried up to bed without ever stirring. This is how it is to fall asleep without any concern for the next day, without any worries about money or work, asleep before your head hits the pillow.

See also The best-deserved sleep, p. 84

Where did you get those pyjamas?

PYJAMAS ARE A SARTORIAL LEGACY of the Indian Raj and centuries of European involvement in southern Asia. The colonial British brought back many words from India, such as 'bungalow', 'pundit' and 'caravan'. 'Pyjamas' were part of this cultural luggage, the word meaning a garment worn over the legs. In the later years of the nineteenth century, these new-fangled pyjamas became a fashionable alternative for men to wear at night, rather than long nightshirts. In the 1920s, high-society flappers began wearing them too, popularising pyjamas for women.

For a flavour of how pyjamas were first received, it's worth quoting that idiosyncratic Edwardian dictionary *Colloquial Anglo-Indian Words and Phrases,* 1903, compiled by Henry Yule and better known as the Hobson-Jobson dictionary.

PYJAMMAS, s. Hind. *pāē-jāma* (see JAMMA), lit. 'leg-clothing.' A pair of loose drawers or trowsers, tied round the waist. Such a garment is used by various persons in India, *e.g.* by women of various classes, by Sikh men, and by most Mahommedans of both sexes. It was adopted from the Mahommedans by Europeans as an article of *dishabille* and of night attire, and is synonymous with Long Drawers, Shulwáurs, and Mogul breeches. It is probable that we English took the habit like a good many others from the Portuguese. Thus Pyrard (c. 1610) says, in speaking of Goa Hospital: 'Ils ont force *calsons* sans quoy ne couchent iamais les Portugais des Indes' (ii. p. 11; [Hak. Soc. ii. 9]). The word is now used in London shops. A friend furnishes the following reminiscence: 'The late Mr. B—, tailor in Jermyn Street, some 40 years ago, in reply to a question why pyjammas had feet sewn on to them (as was sometimes the case with those furnished by London outfitters) answered: "I believe, Sir, it is because of the White Ants!"'

See also The unmanly pillow, p. 24

Sleeper trains: red carpets and killers

EVEN THE IDEA OF SLEEPER TRAINS is something to savour. It conjures up evocative images of 1930s travel posters, the sharp angles and lush colours of a train about to disappear into the night. These were trains that were going to cross continents in a rush of steam, sophisticated couples drinking in chintzy splendour, probably someone being shot in one of the sleeping compartments by a mysterious double-agent who was being pursued by a stylish collection of detectives and thriller writers. After the killer had been thrown into custody by a sharply dressed police chief, everyone else could return for a long cool drink in the bar.

zzzzzzzzzzzzzzzzz

> *Sleeper trains originated in the United States, the need for such sleeping carriages created by the sheer length of the journeys. But somewhere along the way they acquired a romantic, stylish, rather adventurous identity.*

Sleeper trains originated in the United States, the need for such sleeping carriages created by the sheer length of the journeys. But somewhere along the way they acquired a romantic, stylish, rather adventurous identity. It was where strangers were thrown together, where people might be trying to escape, where different cultures travelled side by side.

These were trains whose names had a sense of occasion. Running between New York and Chicago was the luxurious *20th Century Limited,* which entered service in 1902. Passengers had a red carpet waiting for them when they disembarked. The train itself, re-designed in a classic art deco style in the 1930s, had its own trademark cocktail, an on-board barber and a carriage that could be used as a bridal suite.

When you think of overcrowded air terminals and their long queues and lookalike shops, it doesn't really compare to the romance of long-distance rail trips, where the drowsy passenger could look through the window and catch a glimpse of the Russian forests or the peaks of the Pyrenees. Before the First World War, the *Nord Express* sleeper train pulled out of Paris and snaked across the continent to St Petersburg, the *Sud Express* headed southwards from Paris, down through Spain to Portugal. On the *Orient Express* there were spies, lovers and millionaires travelling from Paris to

Night trains: style, speed and romance.

Budapest and Vienna and on to the distant shores of the Black Sea. *Le Train Bleu* took playboys and flappers in the 1920s from Calais and Paris down to the beaches and hotels of the French Riviera.

There are still two sleeper train services in Britain: the Caledonian Sleeper, which runs between London Euston and a number of Scottish stations; and the Night Riviera, which operates between London Paddington and Cornwall.

The names conjure up W. H. Auden's 'Night Train' poem, written in 1935. 'This is the Night Mail crossing the border, Bringing the cheque and the postal order.' And the romance is reminiscent of the classic strangers on a train scene from *North by Northwest* when Cary Grant and Eva Marie Saint begin to kiss.

EVE: You know this is ridiculous. You know that, don't you?
ROGER: Yes.
EVE: I mean, we've hardly met.
ROGER: That's right.
EVE: How do I know you aren't a murderer?
ROGER: You don't.
EVE: Maybe you're planning to murder me right here tonight?
ROGER: Shall I?
EVE: Please do. [*Another long kiss.*]
ROGER: Beats flying, doesn't it?

See also Counting sheep, p. 162

'I must have dozed off': icons of sleep

Rip Van Winkle, celebrity slacker

You've had a couple of drinks, it's a lovely day, a convenient tree provides some shade – next thing you know, you wake up and it's twenty years later, your wife has died, your daughter has grown up and there's been a revolution. It's happened to us all at some stage. Mr Van Winkle's motivational tale, written by Washington Irving, was published in 1819 in a short-story collection called *The Sketch Book of Geoffrey Crayon, Gent.* Rip Van Winkle is still a household name, but who has ever heard of Geoffrey Crayon? I rest my case.

Dylan, musical rabbit

The background is a psychedelic canvas of fluorescent leaves and strange white spaces. There's a guy on a spring with a moustache, a talking cow and a dog with the personality of a rather melancholy middle-aged man. Amidst all this a guitar-strumming figure is leaning against a tree. It's the hippy rabbit, Dylan, in *The Magic Roundabout.* Was it my childish imagination, or was he really always about to fall asleep, those long colourful ears folding gently over his eyes? When I think of falling asleep outdoors on a sunny afternoon, I think of Dylan dozing in his multi-coloured garden, forever on the cusp of oblivion, before the early evening news.

Divine rest: St Vitus, patron saint of oversleeping.

Private Godfrey, Home Guard

There's a theory that in any workplace you can find the characters from *Dad's Army*. There's always a pompous, self-made, vulnerable Mainwaring, a diffident Wilson, an over-eager Jones, a spivvish Walker and a callow Pike. But for the hero of sleep, the Stakhanov of slumber, we have to turn to Private Charles Godfrey. This rose-cottaged warrior knew the value of an afternoon nap. Let's salute his autumnal drowsiness, the decency of the soporific soldier. Let's join him in his catchphrase: 'I must have dozed off.'

Freda, children's television tortoise

Who was the longest-running performer appearing on *Blue Peter*? Freda the tortoise, who appeared on the era-defining children's programme from 1963 to 1979, outlasting John Noakes, Valerie Singleton, Lesley Judd and Petra the dog. Freda was originally introduced as Fred, because it was thought that she was a male. Her greatest contribution to the programme was falling asleep in public. Every year this icon of sleep was put into a cardboard box in preparation for her journey into hibernation. Following in Freda's shell-shaped shadow were two tortoises, Maggie and Jim, two long-forgotten heroes of the show who died tragically in a cold snap in 1982. Lest we forget.

St Vitus, patron saint of oversleeping

Want to stop oversleeping? This fourth-century Sicilian martyr is the patron saint who can be called upon to prevent such a bad start to the day. His connection with oversleeping comes from a rooster that was thrown into boiling oil with the saint. The link with the rooster's early-morning call evolved into a belief that this saint could prevent oversleeping. St Vitus is also the patron saint of comedians. Funny that.

See also Heroic sleepers, p. 65

The unmanly pillow

FOR SOMETHING SO USEFUL, the pillow gets very little appreciation. In fact, it's more likely to be treated with suspicion. Think of the phrase 'pillow talk', with its suggestion of loose talk and loose morals.

This doubt about pillows goes back a long way. The Elizabethan writer William Harrison attacked the namby-pamby men of the 1570s who had become so soft that they were putting pillows under their unmanly heads. In his father's day, a pillow was only used by women who were about to give birth.

Never mind your girly pillows, says Harrison. When he was a youngster, real men used 'a good round log under their heads instead of a bolster or a pillow'. Underneath him would have been 'straw pallets, on rough mats covered only with a sheet, under coverlets made of dagswain or hopharlots'. If he had the luxury of a sack of chaff under his head, he would have 'thought himself to be as well lodged as the lord of the town'.

With distinct echoes of Monty Python's Yorkshiremen, he says such frugal living was still to be found up north, or 'in some parts of Bedfordshire and elsewhere further off from our southern parts'. It was all the fault of those soft southerners, for whom distant Bedfordshire seemed perilously polar. Although, as an aside, 'going to Bedfordshire' is also antique slang for going to bed, so maybe this was part of a geographical pun.

Another type of head rest is the afore-mentioned bolster, a long thin cushion that could spread across the full width of the bed. Years ago you could still find the occasional bolster lurking in old-fashioned houses and hotels, but considering its lumpy discomfort it's no surprise that it's been replaced by the single pillow.

z z z z z z z z z z z z z z z z z

There are pillows with music players inside them, there are scented pillows in the shape of fruit, a pillow that plays the sound of a beating heart and pillows with soft lighting inside.

In the Far East the bolster still survives, where it is regarded as a reassuring comfort in the bed, particularly for children. In China it's known as a 'hugging pillow', and it was traditional for travelling men to take bolsters with them on their journeys. This gave the bolster the name of 'Dutch wife', which also has less salubrious connotations.

There is also a 'husband pillow', which has supportive arms – like or unlike a real husband, depending on your point of view. There are rather sinister-looking versions of these with giant hands. They look as though Postman Pat has been trapped inside a cushion cover. In Japan there are also 'lap pillows', which are pillows shaped like curvaceous ladies' laps. Enough said.

All kinds of unlikely gadget pillows also exist. There are pillows with music players inside them, there are scented pillows in the shape of fruit, a pillow that plays the sound of a beating heart and pillows with soft lighting inside.

As a tribute to the Mafia-related phrase about not

wanting to wake up with a horse's head in your bed, a horse's head-shaped pillow ('stuffed with non-allergenic soft polyester') is available. It might be a death threat, but at least you won't be sneezing.

Pillow fighting has also moved on from the simple days of children slogging away at each other in a bedroom. It's become some kind of weird internet viral activity, with 'flash mobs' turning up to have pillow fights and take photographs of each other doing it and then posting pictures on websites. There are pillow-fight clubs in places such as France and Switzerland, where pillow fighting is known by the exotic name of '*la pillow-fight*'. In Rome, fights between smartly dressed opponents are recorded in moody black-and-white photography. Only the Italians could manage to make pillow fighting look incredibly stylish, with not a feather out of place.

While in Italy they're making art out of pillow fights, in Canada and the United States they're making money, with a Pillow Fight League. This spectator sport features pillow fights between women with names such as Olivia Neutron-Bomb and Dinah Mite. The women contestants belt each other energetically with pillows in front of a noisy crowd until someone is declared the winner. There are more tattoos on show than at an old sailors' reunion. What would they have said at St Trinian's?

See also Bed-in protest, p. 86

How to put on a quilt cover

YOU COULD PROBABLY learn all you need to know about post-war history from the phrase 'continental quilt'.

The austerity years and even later into the Swinging Sixties continued to see the dominance of scratchy layers of sheets, blankets and a bedcover. It was time-consuming to make a bed, a chore that depended on someone being at home and willing to do this in every bedroom each morning.

Enter the continental quilt, carrying the exotic aroma of European holidays and a brave new world of convenience. Instant mash and Angel Delight were in the kitchen; the continental quilt was in the bedroom. No more laborious bed-making, it was now a quick shake and the quilt was back in place. The bedroom would never be the same again.

These early quilts made a virtue of their modern materials. While blankets were made from old-fashioned wool, continental quilts were stuffed with obscurely scientific modern synthetic fibres, probably the shavings from astronauts' jumpers or something similar. The fact that they might explode in the vicinity of a cigarette lighter was only part of their edgy modern appeal.

Quilts have conquered all before them and are now the dominant bedding across the western world. They came, they saw, they covered. The 'continental' tag fell off along the way and the stuffings are now energetically organic, promising to contain the feathers of all kinds of wild creatures.

But one problem has persisted. How do you put the quilt

inside the cover? How do you avoid looking like a drunk wrestling with a sail?

The answer is here: Turn the quilt cover inside out and then put your arms inside this reversed quilt as far as they will go and grab the two furthest corners. With your arms still inside the cover, take the two corners of the quilt (I think of the corners as ears for some reason) in your hands, so that the quilt corners are gripped from within the quilt cover. This is where the pleasurable bit begins. Holding on to the quilt corners, shake the quilt and its cover vigorously. Keep doing this and the cover will unfurl onto the quilt, the right way out. Soon the quilt cover will have been shaken into place over the full length of the quilt.

This really shouldn't be as satisfying as it is.

See also How the ancestors slept, p. 109

The Big Sleep: *the best sleep movies*

THERE'S A THEORY that part of the subconscious appeal of movies is that watching them is like dreaming. The watcher sits silently in a darkened room, absorbed by the images and sounds being screened, losing themselves in the story.

Movies have also always had a soft spot for dream sequences, using them as a way of shifting the plot backwards in time or playing out something that would never happen in reality. Sleeping and dreaming have become part of the language of cinema.

So which are the best sleeping movies? In no particular

order, here are ten films with a link to dreaming or sleep. It is an arbitrary and entirely unscientific list, drawn up without the benefit of focus groups or phone polls, and all the better for it.

The Big Sleep

This smoky, smouldering 1946 classic, starring Bogart and Bacall, is one of the best-known film noir detective movies. The plot is about blackmail, beautiful women and murder, but the title makes one of the oldest literary connections – sleep and death. It goes straight back to the Greek legends, in which sleep and death were twin brothers. The big sleep is the long rest of the dead.

Groundhog Day

Every morning the radio-alarm kicks in at exactly the same point, with the same song (Sonny and Cher's 'I Got You Babe') and the same annoying presenter's patter. Just for that moment alone this 1993 comedy, starring Bill Murray, would qualify as a great sleep movie. Trapped in an endlessly recurring day, the grumpy Murray is taken on a reluctant romantic journey. Is this a dream within a dream, within a dream?

Belle de Jour

It's claimed that Sigmund Freud only ever went to see one movie at the cinema, a not very good cowboy film. But he may as well have written the script for *Belle de Jour*, the 1967 film made by Luis Buñuel and starring Catherine Deneuve. It's about sexual fantasy and dreams and wish fulfilment, repression and release, with a plot about a young wife who becomes a prostitute in the afternoons. Very European, very surreal.

Snow White

This 1937 children's epic, with its enchanted sleep, is the granddaddy of all animated movies. Walt Disney's first full-length feature film showed that cartoons could work at the box office and disproved Mrs Disney's claim that 'No one is going to pay a dime to see a dwarf movie.' The artwork is fabulously rich and soothing, completely different from the modern computer-generated images. Adolf Hitler was an enthusiastic fan of the film. It makes you wonder who he indentified with.

Wayne's World 2

The Jim Morrison dream and the idea that you can go into a dream sequence by waving your arms in a wobbly way like a cartoon character ensure that this 1993 film, written by Mike Myers, qualifies as another major contribution to movie-making about sleep. 'So, did Jim Morrison give you Del Preston's exact address?' 'Yeah, he said exactly London, England.'

Wild Strawberries

This Swedish masterpiece is composed of a series of dreams and flashbacks, in which an old professor looks back on the course of his life. Released in 1957, it was written and directed by Ingmar Bergman, and it's worth leaving the DVD box around the house just to look cultured. The bonus feature is that it's a really excellent movie too.

Crimes and Misdemeanours

This Woody Allen film, released in 1989, is dark and cruel as well as funny. It's not so much rom-com as doom-com, much harsher than Allen fans might usually expect. It's on this list for two reasons: it captures that stormy, restless feeling when you can't sleep; and it has a flashback dream sequence, with the lead character going back to a childhood home, which is a direct reference to Bergman's *Wild Strawberries*. Allen could also have snuck onto this list with *Sleeper*.

Monsters Inc

This takes children's night-time fears of what might leap out in the dark and turns these nightmare creatures into sympathetic characters. They're just ordinary monsters trying to earn a living, pay the bills and keep a cave over their heads. Everyone who bought a DVD player in 2001, the year the film was released, almost inevitably bought a copy of this.

Four Weddings and a Funeral

This contains one of the best depictions of that moment when you wake up and realise that you're already late for something important happening a long way away. If there had been a separate Oscar category for movies about oversleeping – and frankly it would be more interesting than many of the categories – this 1994 comedy would have been in with a decent shout. Well, maybe not a shout, but as the Partridge Family might have muttered . . .

Wizard of Oz

There are three irresistible reasons why this 1939 fantasy has to be included. First, the whole Land of Oz is an extended dream sequence and as weird as a box of drug-addled frogs. Second, anyone who has ever fallen asleep on a hot afternoon can empathise with the scene where Dorothy can't keep her eyes open in the poppy field. It really catches that tempting urge to sleep. Third, it's been on television so often that watching it is like being in that field of poppies: you know you shouldn't fall asleep, but it's so comfortable . . . next thing you know there are flying monkeys everywhere.

Taxi Driver

That's right, this is a top ten. And in at number eleven is that washed-up, washed-out, up-all-night classic, with Robert de Niro as the insomniac cabbie, hungrily searching for something more than what he sees each night on the trashy streets of New York. This is insomnia squared, dysfunctional and lost. And maybe it's a coincidence, but the taxi-driver is called Travis, the name of another great exhausted, sleep-starved movie character, the lost brother played by Harry Dean Stanton in *Paris, Texas*. 'You don't sleep much, do you?'

Even if there were a hundred movies in this top ten, there still wouldn't be room for *Sleepless in Seattle*.

See also The joy of diagonal sleeping, p. 133

Forty winks

GETTING FORTY WINKS means taking a short nap, but why forty and why winking? As a colloquial phrase, it's been around since at least the nineteenth century. George Eliot's novel *Felix Holt,* published in 1866, includes a description of a character lying down and 'having "forty winks" on the sofa in the library'.

Although 'winks' is now used to mean shutting the eyes briefly to send a signal, its origin is from an old English word that meant to close the eyes. It's related to the modern word 'wince', another involuntary form of shutting the eyes. By the Middle Ages, this meaning had changed so that 'winken' meant to close the eyes to sleep. A Middle English dictionary defines it as a 'period of sleep, a nap, a fit of drowsiness'. This meaning can still be found in the phrase 'I didn't get a wink last night'.

This still doesn't explain the connection with forty. The most likely origin is that forty was used as a general term to signify a large number. In the Bible the number forty is used repeatedly in this way: the floods that lifted Noah's boat lasted for forty days and forty nights, Jesus spent forty days and forty nights in the wilderness. The number is used in the same way in old stories, such as 'Ali Baba and the Forty Thieves'. Rather than a precise measurement, it's a symbolic way of saying 'plenty'.

So the 'forty' means a large amount and 'winks' means to have closed eyes – or, putting it another way, plenty of shut-eye.

See also Sleeping like a top, p. 56

Bedhead

THIS IS THE TECHNICAL TERM for the state of tousled confusion that afflicts a person suddenly dragged from their bed. The hair looks like a cornfield that has been hit by a tornado. Everything that should be standing up is lying down, everything that should be lying down is standing up. This uncombed state is often indicative of an internal state of bewilderment, usually associated with oversleeping or the partygoer's phenomenon of 'weekend jet lag'.

In later life, when the partygoer becomes a parent, they might be afflicted by a much more physically painful condition known as 'bunk bed head'. This injury happens when the parent buys a bunk bed for their children. The parent bends down to tuck in their loved one, they give a goodnight kiss, and then in a state of gentle forgetfulness, they stand up straight and whack their head very hard and painfully on the bunk bed.

This parental injury is also related to 'scooter ankle', in which a parent carries a child's metal scooter and fails to observe the scooter swinging round in a perfect arc, where the sharp metal edge collides with the ankle. Another similar medical condition is 'Lego foot', where the barefooted parent suddenly discovers the whereabouts of the missing brick. In all cases there are great acclamations of surprise.

See also Weekend jet lag, p. 38

Sleep concerts

WHENEVER I SIT DOWN in a cinema, regardless of how much I'm looking forward to the film, it takes a huge effort not to drift off to sleep. It's warm, dark, the chairs are deep and comfortable, no one is talking to me, the mobile is switched off, the conditions are perfect for a private screening of *The Big Sleep*. The eyes close and the next thing you remember is a dig in the ribs as you've started to snore like a chainsaw in a metal bucket.

The Japanese, ever creative, have come up with an event that answers all these desires. It's a sleep concert, entirely dedicated to helping stressed workers have a decent chance of some shut-eye. People pay for a ticket, make themselves comfortable in reclining seats, listen to the music – and then do their best to fall asleep. There are some great photos of rows of dozing concert-goers, delighted at getting the social permission to crash out and catch up on some sleep.

This idea has also been exported to Spain, with the concept of 'healing music' being played to an audience encouraged to fall asleep at the first opportunity. The music is meant to summon up the sense of looking up at the stars or hearing the rustling of the wind. More likely it's going to summon up the sound of lots of people sighing with relief that they can fall asleep with a clear conscience, instead of pretending to enjoy the performance.

There are CDs of the music from these sleep concerts, but without wanting to knock a marvellous idea, if they're any good and they work, then how do you ever get to hear how they finish? Should we read anything into the title of one, *Silence*? Does the band quietly down tools when everyone has fallen asleep, go to the bar for a swift one and then return for a sudden show of activity at the end when everyone has to wake up and go home? How do you know they were playing at all?

See also Sleeping to remember, p. 252

Bed testers

GRAHAM BUTTERFIELD's buttocks are insured for a million pounds. There can't be many Lancashire grandfathers who can say that. But Mr Butterfield's claim to greatness is that he is very good at testing beds. He works at the Silentnight factory in Barnoldswick, where his sensitive rear checks the quality of beds.

Bed tester sounds like one of those fantasy slacker jobs, like chocolate taster or coronation-day flag seller. But Mr Butterfield's bottom has got a serious job to do. 'It may sound ridiculous, but my bottom really isn't like any other,' said the bed-bouncing Mr Butterfield. The textures, the fillings, the softness and support all have to be gauged against his delicate instrument.

Testing beds is something that we should all do when buying a new bed, but according to the Sleep Council, four out of five customers are too embarrassed to try them out in the showroom. There are four million beds sold each year in Britain and most of them are bought unbounced.

This shyness is an unfounded anxiety, says the Sleep Council, because display models are protected against dirty clothes and shoes. It also recommends testing beds jointly with a sleeping partner and to try each bed for at least fifteen minutes. This is all beginning to sound a little uncomfortable. You would need nerves of steel to lie down for a full fifteen minutes in a busy showroom.

There is also advice from the Sleep Council on what constitutes the right amount of support in a bed. 'Lying on your back, place your hand in the small of you back and then try to move it about. If it moves too easily, the bed is too hard for you; if it's a struggle to move it, then the bed is too soft. If you can move your hand with just a little resistance, then the bed is just right for you.'

There are also volunteers who test hotel beds, not for their comfort, but for their properties of springiness. There are websites on which people publish photographs of themselves jumping on hotel beds all over the world. Bored businessmen, hyperactive families and energetic conference-goers are all pictured bouncing on their hotel beds. This isn't jumping up and down in a conventional sense, but bouncing from a laying-down position so that it looks as though the person in the photograph is floating or levitating above the bed, or flying horizontally like Superman.

Why? You've got into your hotel room. You've checked the television, the minibar, the bathroom. What else can you do? Jump on the bed. It's rather like the old Billy Connolly line: 'Never trust a man who when left alone with a tea-cosy doesn't try it on his head.'

See also Bed-in protest, p. 86

Weekend jet lag

DON'T EXPECT ANY SYMPATHY, because this is payback for having too good a time at the weekend. When very late nights on Friday and Saturday turn the body's sleeping patterns upside down, and the body thinks it's been spending the weekend in another time zone, Monday morning brings a serious dose of 'weekend jet lag'. The alarm clock might be ringing, but the body doesn't know if it's breakfast or bedtime.

This is most likely to affect youthful clubbers, and another concern for teenagers is that they are not getting enough good-quality sleep during the week, either. The label 'junk sleep' has been used to describe the interrupted, restless sleep of teenagers who never want to turn off their televisions and computer games.

A Sleep Council survey in 2007 found that a quarter of British teenagers were going to sleep with gadgets still turned on at least once a week. These youngsters, who go to bed clutching the remote control or connected to an iPod, spend the following day suffering from tiredness and irritability. Like eating junk food on the move, these youngsters snatch a few hours' sleep, but never really get enough proper healthy, nutritious rest. They suffer from a poor diet of sleep.

Another threat to teenage sleep is the sleepover. 'Sleepover' incidentally is a serious misnomer. No one sleeps – and for the host family, it never seems to be over.

See also Jet lag, p. 197

Lullaby

EVEN THE SOUND of the word is soothing, describing a song that lulls a baby as it drifts away into sleep. Of course, the baby might still be howling and you might be singing in an increasingly hysterical tone, but the principle of a lullaby is a good one. The repetition, the rhythmic sound and the parents' voices all create a sense of reassurance. It's the sound that accompanies the rocking back and forth of a cradle.

The strange thing is that the lyrics of these songs are often so violent:

> Rockabye baby, in a tree top,
> When the wind blows, the cradle will rock,
> When the bough breaks, the cradle will fall,
> Down will come baby, cradle and all.

Here's another shoot-'em-up lullaby that's been around since at least the eighteenth century:

> Cry Baby Bunting,
> Daddy's gone a hunting,
> Gone to fetch a rabbit skin
> To wrap the Baby Bunting in.
> Cry Baby Bunting.

Another Tarantino-esque number used as a lullaby, reputedly about seventeenth-century religious persecution, is 'Three Blind Mice':

> Three blind mice, three blind mice,
> See how they run, see how they run,

They all ran after the farmer's wife,
Who cut off their tails with a carving knife,
Did you ever see such a thing in your life,
As three blind mice.

Lullabies often contain lots of made-up words, which when repeated help to create this sleep-inducing atmosphere. It's difficult to sing them without sounding like a 1950s children's presenter with an accent so cut-glass that you need to wear protective clothing.

Lavender's blue, dilly dilly,
Lavender's green,
When you are king, dilly dilly,
I shall be queen.

But there is something deeply reassuring about these simple rhymes, something hypnotic that reminds us of childhood, something to do with mood rather than meaning.

Hush little baby, don't say a word,
Mama's gonna buy you a mockingbird.

Who cares what it means, it sounds great.

And if that mockingbird don't sing,
Mama's gonna buy you a diamond ring.

Because lullabies also go on and on.

And if that diamond ring turns brass,
Mama's gonna buy you a looking glass.

And so on and so on. The words have that easy, mellifluous quality. Here's another lullaby from the eighteenth century:

Sleep, baby, sleep,
Down where the woodbines creep,
Be always like the lamb so mild,
A kind and sweet and gentle child.
Sleep, baby, sleep.

Lullabies are part of every culture, and there's currently a project running to collect them from across Europe, recognising them as a type of folklore. A brief look at the titles proves the theory that lullabies, a bit like Eurovision Song Contest entries, are about sounds rather than meaning anything. The Romanian entry is 'Haia Haia, Mica Baia'; the Czechs are offering 'Hullee, Baby'; the Greeks 'Nani, Nani, My Child' and the Turkish entry is 'Hu, Hu, to My Baby'. The British contribution is 'Twinkle, Twinkle, Little Star'.

In Britain, lullabies can be traced back to at least the Middle Ages. They were sometimes a type of Christmas carol, nativity songs sung to the baby Jesus. Think of the Coventry Carol, sung in the fifteenth century: 'Lullay, lullay, thou tiny child.'

But lullabies go back much further than this. More than four thousand years ago, the Sumerians had a lullaby that still sounds familiar today, the eternal wish of a parent for their baby to fall asleep, so that they can get some sleep themselves. That's what makes sleep such a universal experience, across time and cultures.

Sleep come, sleep come,
Sleep come to my son,
Sleep hasten to my son.
Put to sleep his open eyes,

Settle your hand upon his sparkling eyes.
As for his murmuring tongue,
Let the murmuring not spoil his sleep.

See also Sleeping to remember, p. 252

Pepys's erotic dreams

SAMUEL PEPYS must be the only diary writer to have his own catchphrase: 'And so to bed.' This chronicler of everyday life in London in the 1660s was a great enthusiast for the bed. After his long accounts of his day's work and the people he had encountered, time and again he returns to the theme of finishing the day with supper and getting back into bed.

Pepys's attention to detail gives a first-hand view of how our Stuart ancestors slept. For anyone who works late too often and sleeps too little, it's surprisingly familiar. He complains when he has to be up at six in the morning for a seven o'clock meeting, and he talks about his wife not coming to bed until one in the morning and the late hour he gets to bed when he has a long day at the office (which he calls 'the office'). On 19 June 1665 he wrote: 'So thence home and to supper, a while to the office, my head and mind mightily vexed to see the multitude of papers and business before [me] and so little time to do it in. So to bed.' When he gets the opportunity, he has a lie-in to recover. 'Lay long in bed, my head aching with too much thoughts,' he began his entry for 9 June 1665. On a Sunday, Mr and Mrs Pepys might stay in bed until ten or eleven in the morning.

Going to bed was not always a private matter. When

Pepys travelled he sometimes had to share a room with someone else; and when an unmarried couple stayed one night, Samuel shared a bed with the man, while his wife, Elizabeth, slept with the female guest.

Pepys recorded his thoughts in a form of shorthand, and the more delicate the subject matter – such as getting caught in an extremely compromising position with a servant girl – the more elaborate the code in which it's written. As well as keeping his thoughts secret, it feels like avoiding writing directly about something embarrassing. He uses bits of French, Spanish and Italian and his own code words, all jumbled together. A frequently used code word is 'cosa' – an Italian word for 'thing'. He records how a lady friend 'did tocar mi cosa con su mano', which means that she 'touched his thing with her hand'. Pepys took a great deal of interest in 'mi cosa' and he often wrote anxiously about his desires, erotic encounters and fears over their discovery.

The language he uses almost has a schoolboy quality. 'I did come to sit avec Betty Michell, and there had her main, which elle did give me very frankly now, and did hazer whatever I voudrais avec l'', which did plaisir me grandement.'

A year previously, Pepys had written: 'I to St. Margaret's Westminster, and there saw at church my pretty Betty Michell. And thence to the Abbey, and so to Mrs. Martin and there did what je voudrais avec her, both devante and backward, which is also muy bon plazer.' Not much need for a translation there.

This bed-hopping wasn't just casual sex, it often seems quite calculating. Pepys was a naval administrator, and he spent a fair bit of time in the company of the husbands of his lovers, helping them in their careers. He was particularly helpful in getting his lovers' husbands jobs that involved long sea journeys.

But back to the bed. Pepys admitted that there were two things in life he could never resist: women and music. When he went to court he was a keen observer of the beauties of the day, watching the peacocking of the Restoration aristocracy and their hangers-on. These mistresses and favourites of Charles II were not only recorded by Pepys in his diary, but also featured in his sleeping life, for Pepys was an enthusiastic dreamer about women. In his sleep he could enjoy his 'sport' without any sense of guilt or fear of being caught, even with powerful figures such as the king's mistress, Lady Castlemaine:

> Something put my last night's dream into my head, which I think is the best that ever was dreamed – which was, that I had my Lady Castlemayne in my armes and was admitted to use all the dalliance I desired with her, and then dreamed that this could not be awake but that it was only a dream. But that since it was a dream and that I took so much real pleasure in it, what a happy thing it would be, if when we are in our graves (as Shakespeere resembles it), we could dream, and dream but such dreams as this – that then we should not need to be so fearful of death as we are in this plague-time.

Samuel Pepys, the shrewdly observing self-made man, also records 'sporting in my fancy' with the queen and Frances Stuart, the great beauty who eluded the energetic charms of Charles II.

But what of Mrs Pepys? How did she feel when she stumbled across his often rather squalid episodes of unfaithfulness and his irrepressible randiness? With suitable hypocrisy, Samuel Pepys had himself been extremely jealous of the time his wife had spent with a dancing teacher, during which he had listened anxiously to the sounds of their feet moving on the floor upstairs.

Appropriately, when Mrs Pepys decided to escalate one of their many quarrels over his philandering, she did so when her husband was in his favourite place – his bed. At one o'clock in the morning, in a furious temper, she pulled open the curtains on his side of the bed, brandishing a pair of red-hot fire tongs. It's no mystery where she planned to put them. But they managed to make up and a few days later they went to the theatre together, where appropriately they saw *The Tempest.* The relief is evident in his diary entry: 'When we come home, we were good friends; and so to read, and to supper, and so to bed.'

Not long after this entry, Pepys stopped keeping his diary, and the painstaking detail of his daily life ends. However, there has been research into what happened next. The big row that had ended with Mrs Pepys threatening to clip him with the tongs was over his dalliance with a girl called Debs – a servant called Deborah Willet. Debs, who seemed to have somewhat obsessed Pepys, was fired from the household. But years later, Debs contacted the now more influential Samuel Pepys, wanting him to help her family. Of course, he was happy to oblige. Debs had married a theology graduate and Pepys, living not far from them in London, found him a suitable job – as a naval chaplain, which sadly would mean spending much of the year away at sea.

See also Roundheads and Cavaliers, p. 87

Andy Warhol's five-hour Sleep *movie*

There's that awful feeling at the cinema when you realise that a movie is going to be much longer than expected. You start checking your watch and shifting in the seat . . . is everything going to be shut when you get out? Perhaps you should nip out early.

So how about a film that runs for five hours and twenty-one minutes in which nothing happens? No one even says anything. In black and white?

z z z z z z z z z z z z z z z z z

According to reports, after an hour of pretty much nothing happening, someone ran out of their seat in exasperation, went up to the screen and yelled 'Wake up!' into the giant ear of the sleeper. As the film progressed, people who didn't get their money back threatened to trash the cinema.

It could have been even longer. When the idea was first conceived, the artist Andy Warhol planned a film that would be as long as a night's sleep. It would run for eight hours, showing nothing apart from someone asleep. There would be no story, only the sights and sounds of someone sleeping, hour after hour.

The original idea was to have Brigitte Bardot as the star, but in the end it featured the poet John Giorno, who was recorded sleeping in the summer of 1963, using a single static camera. *Sleep* was released in January 1964.

At its initial screening in New York, only nine people turned up and two left after an hour. When it was screened in Los Angeles, there were 500 people in the audience, with 50 staying awake until to the end. According to reports, after an hour of pretty much nothing happening, someone ran out of their seat in exasperation, went up to the screen and yelled 'Wake up!' into the giant ear of the sleeper. As the film progressed, people who didn't get their money back threatened to trash the cinema.

Why would anyone want to make such a film? Warhol explained that he wanted to catch sleep on camera before it became 'obsolete'.

> I could never finally figure out if more things happened in the Sixties because there was more awake time for them to happen in (since so many people were on amphetamine), or if people started taking amphetamine because there were so many things to do that they needed to have more awake time to do them in . . . Seeing everybody so up all the time made me think that sleep was becoming pretty obsolete, so I decided I'd better quickly do a movie of a person sleeping.

See also Surrealism and dreams, p. 235

Where is the Land of Nod?

THIS PHRASE has a dark, bloodstained origin, the Land of Nod being the biblical location where Cain fled after he had killed his brother. It was located 'east of Eden', which gave its name to the John Steinbeck novel and James Dean movie.

The first recorded use of this phrase as a humorous way of talking about sleep was by the satirist Jonathan Swift. In his *Compleat Collection of Genteel and Ingenious Conversation* (1738), Swift had them rolling in the aisles with this exchange: 'I'm going to the Land of Nod. Faith, I'm for Bedfordshire.' In the Bible, the Land of Nod is a bleak and unwelcoming place for the wanderer, but Swift has colonised it for the comfortable realm of sleep. He was also drawing on an existing use of the word 'nod', meaning when the head would nod drowsily into sleep, as in 'nodding off'.

By the nineteenth century it was an established part of the language of childhood, with Robert Louis Stevenson writing a nursery rhyme: 'From breakfast on through all the day, / At home among my friends I stay, / But every night I go abroad, / Afar into the land of Nod.' No references at all to fratricidal killers. The Land of Nod had become part of the Victorian landscape of sleep, with nightshirts and those pointy nightcaps. It was the kind of place where Wee Willie Winkie, a character from an 1840s children poem, might have been living.

There is another real-life Land of Nod. It's a place in East Yorkshire and you can stand next to a road sign pointing in its direction. Mind you, Yorkshire seems to specialise in quirky names. Is there really a village called Wham? In a similar vein, the phrase 'Land of Nod' is not to be confused with the Land of Noddy, which is the 1950s Home Counties, or the Land of Noddy Holder, which is of course Birmingham.

See also Caught napping, p. 123

Fairy-tale ending

THERE IS SOMETHING MAGICAL and rather eerie about sleep. The sleeper is present, but at the same time absent. They might be in the same physical place, lying in the bed looking vacant, but they've entered a world that no one else can see. So it's not surprising that sleeping and

Fairy tales are filled with dreams, enchanted sleeps and strangers in the bed.

dreaming have been such evergreen themes in fairy tales and legends. Look at the sleeper and it's not difficult to think of sleep as an enchantment or a magic spell, keeping the sleeper in a state of mysterious suspension.

'Sleeping Beauty', a story with roots reaching back into medieval Europe and probably much further, has that classic fairy-tale mixture of threatened innocence, romance, violence, rescue and redemption. It's the type of universal tale that no one can resist trying to interpret, whether it's Walt Disney or psychoanalysts. Fairy tales always have that dream-like, bizarre quality. A beautiful young girl lives under the threat of a curse that eventually strikes her. The girl remains in a deep sleep for a hundred years, and all around her the courtiers and guards fall into this same frozen slumber, hidden away in a castle that becomes more and more densely wrapped in a cloak of thorns and brambles. The girl is saved by a heroic prince who penetrates this forest, but who is terrified to find the 'image of death everywhere' in the faces of the silent bodies around the castle. As ever with fairy tales, darkness is never very far below the surface. But the prince soon realises that these people are not dead but sleeping, and when he kisses the beautiful girl, he breaks the spell.

zzzzzzzzzzzzzzzzzz

It seems like the weirdest dream that anyone ever had – dwarf houses, glass coffins and magic sleeps. But it has had a perennial appeal for film-makers.

The story first reached a wide audience at the end of the seventeenth century in a compilation of stories by the French writer Charles Perrault. In his version there is an even more savage twist, when the rescued girl is brought back to the prince's house. The prince's mother is a wicked old queen who tries to eat the girl and her children, until once again the prince makes a timely intervention. Somehow or other the cannibalism theme never made it into the Disney film.

For psychologists and psychoanalysts such stories are a treasure trove. The enchanted sleep has been interpreted as a symbol of latent sexuality, a young woman stifled and silenced, trapped behind a wall of thorns, until eventually a partner ends her long wait. In Perrault's version, the prince doesn't hack his way through the undergrowth, it parts to let him in.

Snow White is another beautiful girl who gets to oversleep. A wickedly jealous queen gives the girl a poisoned apple that sends her into a deep sleep. The broken-hearted seven dwarves with whom she has been staying put her body into a glass coffin. Here she stays, unable to wake, until a passing prince falls in love and brings her back to life. In the original version the wicked queen is forced to wear a pair of red-hot iron shoes as a punishment.

It seems like the weirdest dream that anyone ever had – dwarf houses, glass coffins and magic sleeps. But it has had a perennial appeal for film-makers. Long before the Disney cartoon, there was a silent version of Snow White. Can you imagine what a hallucinogenic night out that would have seemed?

People can't seem to stay awake in fairy tales. Goldilocks walks into a house owned by a family of bears and five minutes later she lies down and falls asleep. It's only a quick getaway that saves her from the wrath of the house-proud occupants. Maybe this makes more sense in the early English versions of the story, which feature an old vagrant woman rather than a young girl. In fact, it becomes a much sadder and hungrier tale of a homeless woman looking for food and shelter. When the story was rewritten the old woman was replaced with a youngster called Silverlocks, and in the twentieth century, this character was renamed Goldilocks.

A bed also plays an important role in 'Little Red Riding Hood', with the big bad wolf using the granny-in-the-bed routine to trap his victim. This has been interpreted as a story about a girl's coming of age, the bed being the symbolic place where she changes identity and becomes an adult.

The wolf didn't have any complaints about the lumpiness of the bed, but a much more delicate approach to sleep appears in 'The Princess and the Pea'. A pea is placed below twenty mattresses and twenty feather beds, but the princess is such a finely bred sleeper that this is enough to keep her awake all night. In the kind of logic that only makes sense in dreams and fairy tales, this pea-disrupted sleep makes the princess completely irresistible to the prince, who immediately asks her to marry him. Clearly, being kept awake by small vegetables was an unmistakable sign of sophistication. The princess had passed the sleep test. She might have needed a ladder to get on top of the tower of mattresses and she was going to be hell to sleep next to, but there was no doubting her princess credentials.

For all the princesses there was a fairy-tale ending. They slept happily ever after.

See also Freud and Jung, p. 224

Best-ever joke about sleep

HOW MANY HOURS of decent sleep have been squandered by waiting to see what's going to be number one in a television top fifty list, only for this to be a disappointment? You go to bed hours later than you'd intended, feeling cheated. So I'm going straight to the top: here's the best-ever joke about sleep, attributed to Bob Monkhouse. No commentary, no explanations:

> I'd like to die peacefully in my sleep like
> my father. Not screaming and terrified
> like his passengers.

See also Sleep and death, p. 254

Sleep laureates

IF THERE WAS A POET LAUREATE of sleep, John Keats would be dozing at the front of the queue. A leading light of the Romantic movement, he was dogged by ill health, an unhappy personal life and attacks by critics. Completing the perfect poet's CV, he died in his twenties in Rome.

But Keats did love his sleep, and left the most eloquently lush descriptions of the sensations of falling asleep. The sounds of his words with their drowsy, hypnotic softness, are a pleasure in themselves. Sit back and let them wash over you:

'To Sleep'

Soft embalmer of the still midnight!
Shutting with careful fingers and benign
Our gloom-pleased eyes, embower'd from the light,
Enshaded in forgetfulness divine;
O soothest Sleep! if so it please thee, close,
In midst of this thine hymn, my willing eyes,
Or wait the amen, ere thy poppy throws
Around my bed its lulling charities;
Then save me, or the passèd day will shine
Upon my pillow, breeding many woes;
Save me from curious conscience, that still lords
Its strength for darkness, burrowing like a mole;
Turn the key deftly in the oilèd wards,
And seal the hushèd casket of my soul.

Shakespeare might also have made a strong claim to be the poet of sleep, with familiar lines from *The Tempest* such as:

> We are such stuff
> As dreams are made on, and our little life
> Is rounded with a sleep.

Or in *Macbeth,* the best-known play about insomnia:

> Methought I heard a voice cry 'Sleep no more!
> Macbeth does murder sleep', the innocent sleep,
> Sleep that knits up the ravell'd sleeve of care,
> The death of each day's life, sore labour's bath,
> Balm of hurt minds, great nature's second course,
> Chief nourisher in life's feast.

Sleep also appears in Shakespeare's sonnets:

> Weary with toil, I haste me to my bed,
> The dear repose for limbs with travel tired;
> But then begins a journey in my head,
> To work my mind, when body's work's expired:
> For then my thoughts, from far where I abide,
> Intend a zealous pilgrimage to thee,
> And keep my drooping eyelids open wide,
> Looking on darkness which the blind do see
> Save that my soul's imaginary sight
> Presents thy shadow to my sightless view,
> Which, like a jewel hung in ghastly night,
> Makes black night beauteous and her old face new.
> Lo! thus, by day my limbs, by night my mind,
> For thee and for myself no quiet find.

On a long cold night or on a long, hard journey, sleep is

also a vision of comfort, a hunger for a place by the fire. It's a mood caught in the Robert Frost poem 'Stopping By Woods on a Snowy Evening':

> The woods are lovely, dark and deep,
> But I have promises to keep,
> And miles to go before I sleep,
> And miles to go before I sleep.

You can almost feel your eyelids wanting to close in sympathy. But for a beautifully nostalgic depiction of falling asleep, capturing that poignant, pleasurable sensation of letting go, it's worth turning to Jerome K. Jerome's *Three Men in a Boat* (1889):

> Harris said he didn't think George ought to do anything that would have a tendency to make him sleepier than he always was, as it might be dangerous. He said he didn't very well understand how George was going to sleep any more than he did now, seeing that there were only twenty-four hours in each day, summer and winter alike; but thought that if he did sleep any more, he might just as well be dead, and so save his board and lodging.

There is an irresistible, reluctant poetry about these characters in the boat, gently ambling up the river. Sleep is never far away; it's something rather haunting and sad. The following description of falling asleep in the evening is a lights-out for the Victorian era:

> We, common-place, everyday young men enough, feel strangely full of thoughts, half sad, half sweet, and do not care or want to speak – till we laugh, and, rising, knock the ashes from our burnt-out pipes, and say 'Good-night,' and,

lulled by the lapping water and the rustling trees, we fall asleep beneath the great, still stars, and dream that the world is young again.

See also Dream poetry, p. 237

Sleeping like a top

WHY WOULD ANYONE want to sleep like a spinning toy? It's a phrase that's been in circulation since at least the seventeenth century, when it appears in *The Old Batchelor*, a play by William Congreve: 'Should he seem to rouse, 'tis but well lashing him, and he will sleep like a Top.'

There is a theory that the phrase refers to the apparent stillness of the top's axis when it's in motion. It seems to be completely at rest, i.e. sleeping, when it's revolving at its fastest rate. This ties in with the way the word 'sleeping' is used by modern-day children playing with yo-yos. When the yo-yo is spinning fast at the end of its uncoiled string, this is known as 'sleeping'.

But even more obscure is the origin of 'sleeping like a log', which supposedly derives from the way snoring can sound like the sawing of logs. Although, once again, there might be another origin. Before the days of satellites and electronic navigation, ships used a 'log' to measure their speed. This was a piece of wood attached to a kind of fishing-reel arrangement that was lowered into the water. As the ship moved, the log was left dozing behind in the water and the ship's speed was measured using a series of knots tied in the unwinding reel attached to the log. This nautical use of the word 'log' dates back to at least the 1500s.

There are more disputes over the origin of the phrase to 'sleep tight'. There are claims that it comes from the days when mattresses rested on ropes strung across bed frames, and that to get a good night's sleep the ropes had to be tightened.

But there are word buffs who take issue with this and claim that it's not about changes in bed technology but about the changing way we use words. 'Tight' once meant 'sound', in the sense of a lid being fastened 'soundly'. So sleeping tight would be the same as sleeping sound, held secure in sleep.

See also Counting sheep, p. 162

Futons: beds of torture

FOR ANYONE who has never had the experience of sleeping at ankle height on a piece of bedding about as thick and comfortable as a sheet of metal, a futon is a low flat bed. It originated in Japan, but moved into the *Guardian*-reading parts of Britain some time in the 1980s, particularly in small flats in fashionable London postcodes.

The selling points for the futon are that it looks rather cleverly minimalist and is good for your back. Sleeping on this bone-hard mattress is a kind of tough love for your posture. What seems more likely to me is that this plays to a deep-seated Puritanism that believes if something is really uncomfortable and without pleasure, it must be good for you. Like extreme diets or exercise crazes, the fact that it is joyless becomes a virtue.

I've woken up on futons and felt as though I'd spent the entire night having my back expertly kicked with hobnail boots, every inch of every muscle aching. It can be more like torture than bedding. Even greater pain can accompany the attempt to get up. You stare plaintively sideways, unable to move, your face ridiculously close to the carpet. All you can do is roll off the bed and writhe around in agony on the floor, like a dying fish slapping around on the deck of a trawler.

They might have been fashionable, they might have had a certain exotic student chic, they might have been handy in small spaces, but don't mistake the futon for a comfortable night's sleep. There are American versions which are really just fold-out sofa beds, but if you see the real thing be ready to recognise the enemy. It will look like a wooden pallet covered with a mattress about the thickness of a handkerchief. It is no friend of the long lie-in.

See also Top ten tips for a bad night's sleep, p. 163

Water beds

WHEN YOU THINK of water beds, you think of the 1960s. Austin Powers would have thought that having one was a shagadelic necessity. And appropriately the modern water bed was created in the very epicentre of hippiedom, San Francisco in 1968, following the failure of an earlier project, which was a vinyl chair filled with jelly. It might have been a terrible chair but it would have made a great party.

However, the idea for a water bed is much older. The Persians had water beds made from goatskins more than 3,000 years ago. These beds used the same basic principle of letting the sleeper rest on a flexible support of sealed water. They were solid enough to lie on, but soft enough to feel relaxing.

The Victorians were also enthusiasts for water beds, but they saw them as pieces of medical equipment, beneficial for the sleep of invalids. In the 1830s there was a 'hydrostatic bed' on sale that consisted of a rubberised canvas cover stretched over a container of water. The intention was to reduce the risk of bed sores for the long-term bedridden.

In 1873 Sir James Paget, surgeon to Queen Victoria, presented a water bed to St Bartholomew's hospital in London with the aim of improving the comfort of patients and preventing pressure sores.

The first patent for a water bed was issued in the 1880s, to a doctor from Portsmouth, who wanted the patient to be able to float on the surface of this bed, with less pressure than they would experience on an ordinary mattress. Portsmouth people might be made of sturdier stuff than the sun-kissed Californians, but they still had two significant objections to this prototype water bed: it was freezing cold and it leaked.

The water bed market continued to wobble unsuccessfully, with Harrods only managing to sell a few of these beds by mail order in the 1890s. But fast-forward to the psychedelic, sensation-seeking 1960s and the water bed came back with a very different image. Vinyl-covered and filled with heated water, it was built for fun rather than therapy. The bed, known as the Pleasure Pit, was a huge success in the liberated bedrooms of the swinging Sixties.

The undulating surface became the fashionable choice for hippies and playboys. However, there was a twist in the tale. The inventor, Charles Hall, wasn't able to patent his invention because a similar water bed had been described in detail in a science-fiction novel in the 1940s, and so it was already deemed to have been invented.

Nonetheless, with lots of spaced-out celebs seen clambering onto their water beds, he was still able to sell his design, promoted through the 1970s and 1980s as a symbol of the novelty-hungry, hedonistic Californian lifestyle.

There was no great secret about its appeal in the 1960s. As an advertising campaign promised: 'Two things are better on a water bed. One of them is sleep.'

See also The joy of diagonal sleeping, p. 133

Why do children like frightening bedtime stories?

WHEN I WAS ABOUT FIVE OR SIX years old and looking for a story to read in bed, there was one book that always drew me. It was also the one I hated most. My sister's copy of *Hansel and Gretel* filled me with fear. The drawings, a kind of late-1960s cartoon gothic, made me scared to even look at the book. I can still see the distorted, wart-encrusted face of the witch and the sickly-sweet house that the children couldn't resist.

But I kept going back to find that book. Each step of the story was a nightmare, the children taken into the woods, betrayed and rejected by their parents, captured by an evil

Poppies and opium have long been used as sleep remedies.

witch, threatened with being cooked and eaten. The book seemed to embody all the worst terrors I could imagine, a horror story of misfortune and neglect. So why did I keep going back to take it with me under the bedclothes?

It's one of those strange contradictions that children

have a grim fascination for the books and films that scare them most. It's carried over into adulthood by watching scary movies or reading a ghost story in bed.

The psychological theory is that such stories help us to tame our fears – we familiarise ourselves with the threat, drawing its sting, rehearsing our responses. Fairy stories have huge elemental themes – abandonment, betrayal, coming of age, separation and recovery – and through such tales we can begin to confront our worries about the outside world and growing up. It's a kind of psychological play fight, rehearsing the moves that we need to defend ourselves. We're drawn again and again to look at the enemy, tasting fear but in an environment that we know is safe and secure. It's a way of taking control of what makes us most afraid.

Frightening stories and films, the kind that make them scared to turn off the light, have a special appeal to children. You can see them both wanting to watch and hating to watch, struggling with their own curiosity and anxiety, looking away at the frightening moments and then looking back in case they've missed anything. They seem to want to be spooked, daring themselves to dip their toes into such imaginary danger.

An important part of this is that these are fictional versions of fear. A story about real-life dangerous people would just be sinister and disturbing; it wouldn't have any of the protective power of tales about made-up characters and unlikely situations. My parents used to tell me stories about the 'Eight o'clock Man', a character they'd invented to stop me from staying up beyond bedtime. Although I was afraid of what would happen if the Eight o'clock Man caught me up too late, part of the ritual was the implicit understanding that he didn't really exist.

The ghost story at bedtime is like the frightening ride at the funfair. You can enjoy the momentary sensation of fear, safe in the knowledge that at the end of the ride you can walk away. As long as I don't have to look at the pictures of the witch in *Hansel and Gretel* I'll be fine.

See also Sleep training: quack alert, p. 206

Afternoon sleep in Regent's Park, London in 1930.

The Poor Man's Wealth

Heroic sleepers

ANYONE STUCK in the joyless prairies of an open-plan office will have heard the lonesome call of the office martyr. They were working late again last night, checking e-mails when they got home and had to get up even earlier this morning to start the self-flagellating cycle all over again. And sleep? They never get any sleep, they're so tired that they can hardly think straight. They would give anything to sleep, but there's all this work that has to be finished and if they don't do it then who will?

Whenever people face an impossible volume of work or have to make important decisions, one terrible and unproductive habit of thinking is that going without sleep is a sign of strength. Only getting four hours' sleep is seen as a badge of commitment, a sacrifice that reflects the seriousness of the task.

Nothing could be further from the truth. The really heroic action is to look the enemy in the eye, then put on your pyjamas and go to bed. Refusing to give up on sleep is the true signal of courage. Better decisions are made, and better lives are lived, when leaders are getting a decent night's sleep.

Take the example of Winston Churchill in the toughest times of the Second World War. How did he withstand the gruelling pressures? Imagine how much was expected from him, the lack of time, the fraught search for the right decisions, the public appearances, the emotional demands. With bombs falling all around, what did he do? He went to bed in the afternoon.

Speaking after the war to an American journalist, he explained how he managed to survive the extreme pressures:

> 'You must sleep some time between lunch and dinner, and no half-way measures. Take off your clothes and get into bed. That's what I always do. Don't think you will be doing less work because you sleep during the day. That's a foolish notion held by people who have no imagination. You will be able to accomplish more. You get two days in one; well, at least one and a half, I'm sure. When the war started, I had to sleep during the day because that was the only way I could cope with my responsibilities.'

It's worth dwelling on that for a moment. Can you imagine a modern politician having the confidence to admit to sleeping in the afternoon? Can you imagine a press conference where a prime minister honestly says that their response to the current financial or political crisis is to 'take off my clothes and get into bed'?

There's a rather destructive modern political expectation that working round the clock means better results. It gives the impression of determination, even when the outcomes are not particularly impressive. It shows that at least no one has been enjoying themselves when things are going wrong. When he was backed into a corner and fighting for survival, Churchill showed that what mattered was to get the job done, not to make it look like it was being done. And for Churchill it meant sleeping during the day.

If we look at what he says, it's not about a lifestyle

choice or a personal appetite, he recognised the irresistible importance of sleep. Without the sleeper there would be no fighter. 'I had to sleep during the day because that was the only way I could cope with my responsibilities.'

zzzzzzzzzzzzzzzzz

When he was backed into a corner and fighting for survival, Churchill showed that what mattered was to get the job done, not to make it look like it was being done. And for Churchill it meant sleeping during the day.

Churchill was a spectacular night owl, balancing his long afternoon sleeps by holding meetings in the early hours of the morning. There are stories about the exhaustion of the military chiefs summoned to the prime ministerial retreat at Chequers who found that conferences were beginning long after they would usually have gone to bed.

There was nothing of the puritan about Churchill. His attitude to sleep was not only that it was a necessity, but also that it was there to improve the pleasure of the waking hours. Sleep was a sauce to add flavour to the day. 'A man should sleep during the day for another reason. Sleep enables you to be at your best in the evening when you join your wife, family and friends for dinner. That is the time to be at your best, a good dinner, with good wines . . . champagne is very good ... then some brandy, that is the great moment of the day.'

Again, it sounds such a long way from the hyperactive political life of the present day, when 24-hour news demands instant reactions and a constant feeding of often shallow updates. We expect our leaders to be in a state of perpetual exhausted penitence. Think of the firestorm that would follow if a modern-day prime minister announced that he or she were going to sleep all afternoon in order to heighten the enjoyment of the champagne they were going to drink in the evening.

But these were different times. It wasn't just Churchill who saw the value in following instincts. One of Churchill's doctors, Charles Rob, surprised a medical conference in the 1950s when he delivered a paper on the best practice for treating blood clots. He prescribed whiskey and sleep. 'The best treatment for the condition is rest. The best way to rest is sleep. The best way to get sleep is to relieve pain, and the best way to relieve pain is to give whiskey,' said this former combat surgeon, with irresistible logic. Who could argue with the advice? The surgeon lived until his late eighties and Winston Churchill, drinker, smoker and sleeper, lived until he was ninety.

The idea of heroic sleeping has a long tradition. In fact, there is a particular type of great leader who is distinguished by sleeping deeply at the moment of greatest crisis. It doesn't fit in with our modern obsession with doing more of everything, but great men and women have often been heroic sleepers.

In the sixteenth century the French statesman and writer Michel de Montaigne identified this apparent contradiction – how the great could exhibit a sense of serenity that let sleep pass uninterrupted in the face of danger. It's reminiscent of Hemingway's phrase, quoted by John F. Kennedy, that courage is 'grace under pressure'. 'I have taken notice, as of an extraordinary thing, of some great men, who in the highest enterprises and most important affairs have kept themselves in so settled and serene a calm, as not at all to break their sleep,' wrote Montaigne in his essay on sleep.

As examples, he tells the story of how Alexander the

Great spectacularly overslept on the morning of a great battle against the Persian king Darius, having to be shaken awake. The emperor Augustus also fell asleep just before the start of a naval battle off Sicily, so that he had to be woken to give the signal for attack. It almost seems like a psychological response to extreme tension for these great leaders to retreat into their own inner world of sleep before engaging with the enemy. It also shows a spectacularly steely sense of composure to prepare for a life-or-death struggle by taking a quiet nap. Such ancient examples have a rather uncanny similarity to the image of Churchill sleeping in the afternoon amid the anxiety of the Blitz.

Sleeping in the face of danger can also be a way of instilling confidence in others, showing that there is no need to panic. For an extreme example, how about sleeping when a volcano is erupting, hot ash is falling all around and the foundations are being shaken by tremors? Going back to bed seems like a particularly stoic response, rather reminiscent of the scene in *Carry on Up the Khyber* when the stiff upper-lips of the Raj continue to eat soup as the ceiling falls on their heads. Yet in AD 79, when Mount Vesuvius was exploding, Pliny the Younger recorded how his uncle, Pliny the Elder, reassured his household that there was nothing to worry about. Telling his family that the raging volcano above them was only a minor fire in abandoned cottages, he went to bed. 'Then he retired to rest, and in fact, he relaxed in sleep that was wholly genuine, for his snoring, somewhat deep and loud because of his broad physique, was audible to those patrolling the threshold.'

If snoring could ever be heroic, Pliny's uncle's double-bass snoring is a noble example. The volcano was burning around them while he filled the house with his relaxed snores.

'By this time the courtyard which gave access to his suite of rooms had become so full of ash intermingled with pumice stones that it was piled high. Thus if he had lingered longer in the bedroom the way out would have been barred. So he was wakened, and he emerged to join Pomponianus and the rest, who had stayed awake.'

Not all such stories have happy endings. Pliny's uncle was fully aware of the seriousness of the threat, having been out earlier trying to calm the nerves of frightened locals. Going to sleep was a way of showing that he was not afraid. It was an act of faith in the normal. When he woke, the situation was getting much worse and with tremors shaking down houses, he and his household joined the terrified crowds trying to protect themselves from falling rocks and burning heat. He died, unable to breathe in the choking air, and when his body was discovered two days later, his nephew recorded that he 'looked more asleep than dead'.

See also Heroes under the hill, p. 232

Hot in bed: the electric blanket

ANYONE WHO ENJOYS the warm embrace of an electric blanket at night might be surprised to know its uncomfortable origins: draughty TB clinics and freezing airmen in the Second World War. Bed-warming devices have been around for years in the form of hot-water bottles, ceramic foot-warmers or even hot bricks, and there were some early experiments in the 1900s with using electric currents for warming beds. But the electric blanket really became established in the 1920s as part of the treatment of tuberculosis. Patients who had this killer disease were treated in sanitariums where it was believed they needed lots of fresh air. The tuberculosis sufferers were kept in rooms with the windows open, even at night, which meant that they were often freezing cold. Upmarket establishments countered this by introducing electric blankets, known then as warming pads or heated quilts, to keep the TB patients warm while the fresh air swirled around them. These blankets were rather unwieldy and were not ideal for something that had to be flexible enough to be stretched and rumpled on a bed each night.

z z z z z z z z z z z z z z z z z

There were experiments to design an electrically heated flying suit. From this research came the electric blanket, with electric heating elements running through material in a more flexible, lightweight form

The next stage in the evolution of the modern electric blanket came with research into heated clothing for US airmen during the Second World War. Unheated cockpits were freezing cold and there were experiments to design an electrically heated flying suit. From this research came the electric blanket, with electric heating elements running through material in a more flexible, lightweight form. The first consumer versions went on sale after the war, costing about $40, which meant that they were very much a luxury. The association with wartime pilots also shifted their image: no longer a product for invalids and the elderly, the new blankets were a modern marvel and part of the post-war boom in consumer technology, alongside washing machines and televisions.

For anyone who was a child in the 1960s and 1970s, the electric blanket is a warm memory. By then they had come down in price and were often bought as a cheap way of keeping warm before families could afford central heating. They were part of an era of three-bar fires and thick vests. Cold staircases, halls, landings and bedrooms had a happy ending waiting in the form of the orange glow of the light of the electric blanket. Once you pulled back the covers, there was that warm, slightly singed smell of hot sheets and blankets. No matter how cold it was outside, or if there was still a chill dampness on the pillow, it was snug inside the cocoon of the bedcovers.

It's probably a safety risk, but I found it quite heart-warming to read a report that an electric blanket brought in for checking during one of the recent health-and-safety drives had been received as a wedding present in the 1970s and had been used ever since. The safety officials were appalled, but somehow it seems quite poignant to think of a couple with thirty years of warm nights together using the same blanket.

In modern overheated houses, where we're never really cold, the electric blanket is no longer either a luxury or a necessity. But if I'm lying in bed at night and think back for a moment, I can still remember the reassuring click of the electric blanket switches on my own bed, on my sister's and on my parents' in the bedrooms next door.

See also Water beds, p. 58

If dogs sleep so much, why do they keep yawning?

YAWNING, like many things considered rude in public, is really quite enjoyable. Like a good stretch or scratching an itch, it gives a special kind of hard-to-define pleasure. It puts things back in the right place. There is even a gem of a word – pandiculation – to describe the double-delight of yawning and stretching at the same time.

Despite the common belief that yawning is about increasing the intake of oxygen, recent research has suggested that this is not really the case. It's known that people yawn more when they are tired, bored or extremely nervous, but it's not really clear why such situations should trigger this reflex action. One theory is that it's about cooling the brain, which might help people keep alert under pressure. Others suggest that yawning is part of an arousal signal, warning of tiredness and fending off sleep.

But there is another way of looking at yawning, and that is as a social signal. Yawning really is contagious: when one person yawns, other people in the group will copy; yawning becomes a type of group behaviour. It's rather like the pub game of scratching and then waiting to see how long it is before someone else around the table mirrors this behaviour by scratching themselves. Humans in a group can't stop copying each other, whether it's sitting positions, a turn of phrase or yawning. Experiments in the United States tested this theory by placing someone in a conspicuous position in the middle of a library and then getting them to yawn vigorously. Like wildfire, the yawning rapidly spread around the room.

Humans yawn all their lives: foetuses are yawning in the womb, babies are enthusiastic pandiculators, and the elderly could give masterclasses in complete head-back and teeth-baring yawning. It's also a pleasure shared with the animal kingdom. So-called 'foetal pandiculation' – defined as an 'instinctive movement, consisting in the extension of the legs, the raising and stretching of the arms and throwing back of the head and trunk, accompanied by yawning' – has been observed in unborn lambs.

But back to the central question. Dogs are not remotely sleep-deprived; they sleep whenever they feel like it, often for large stretches of the day. So why are they yawning so much? Because they are copying us. A study from Birkbeck College in London has shown that dogs are influenced by human yawning. In the same way that a human yawner sets off everyone else, so a dog stuck in a room with a yawning human will start to yawn as well – those big slobbery dog yawns that seem to go on for ever. Researchers observed dogs who were with a human who kept yawning, and another control group in which the human did not yawn. The conclusive findings were that the yawning human sparked a wave of yawning dogs, while the dogs in the control group did not yawn at all.

Mind you, if we tried to get all the pandas in the zoo to do the same, would we be pandering to a pandemic of pandiculating penned pandas?

See also Sleep concerts, p. 35

Oversleeping

NOTHING IS AT ONCE so awful and enjoyable as oversleeping. The extra sleep is deeply pleasurable, but you also feel a dreadful sense of terror about how much damage it's going to cause. Even though you're already late for work or you've left it too late to catch a plane, you still rush around as though the panic is going to make up for lost time.

Snooker player Graeme Dott's disastrous trip to a tournament in China in 2002 is a classic case of oversleeping, and was sufficient to earn him an award as the *Guardian*'s Alternative Sports Personality of the Year.

The Scottish snooker player's nightmare trip began with a delayed 43-hour journey from Glasgow to Shanghai, which meant that he arrived a full nineteen hours behind schedule. When he reached the hotel, he set two alarm clocks to wake him in time for his match, but, exhausted and with his body clock confused, he slept through all the attempts to rouse him.

'I remember waking eventually in a blur and seeing that it said 14:15 on the alarm-clock. For a while I just lay there disorientated, wondering what was going on. And then I heard the doorbell. It was so loud. And then I panicked,' he told the *Guardian*.

The snooker player grabbed his clothes and tried to make it to the stadium, in a journey that involved an uncooperative taxi and having to run the last stretch on foot. He hadn't had time even to put on his underpants, and when he arrived the match officials wouldn't let him go to the toilet. Late for the tournament, he was docked two frames. He then lost the match, by those two frames, and had to turn round and fly home again.

'At the time it wasn't funny. It ruined my season really. I'd been going well until that point but I didn't win another match for ages. Everyone started phoning up from radio stations and newspapers wanting to talk about it, but I couldn't. I look back on it now and can smile, but at the time I was devastated.'

See also Sleeping to remember, p. 252

The old enemies: sleep versus work

LEVI HUTCHINS is a name that should strike horror into the hearts of all decent, bed-loving citizens. This eighteenth-century clockmaker in New Hampshire wanted to ensure that he would wake up each morning at 4 a.m. The sun, having more sense, had no intention of getting up at that time to wake him. So the youthful Levi Hutchins devised a clock that could set a loud bell in motion.

In fact, there were probably older alarm clocks. The Germans are thought to have pioneered an automated type of ringing clock a couple of centuries earlier. But it's Levi Hutchins, proud as punch to be up at 4 a.m. with a collection of insomniacs and burglars, who lays claim to be the father of the alarm clock.

The alarm clock is the factory whistle that controls the beginning of the day. There's nothing remotely natural about being wakened this way. Animals in the wild, family pets and bored-looking lions in the zoo wake up when they fancy, and if they feel tired again they roll over and fall asleep. There are no alarm bells in nature shaking animals awake. Sleep is a self-regulating mechanism: if you don't need any more, you wake up; if you still need some, you stay asleep.

zzzzzzzzzzzzzzzzz

Before the Industrial Revolution, people working on the land might have had two stretches of sleep during the night, or else a siesta in the afternoon. The only clock they had to watch was the rising and falling of the sun.

But work, more than anything else, seems determined to disturb our sleep. We've become so accustomed to the way that work shapes our sleeping that it seems normal. It wasn't always that way. Before the Industrial Revolution, people working on the land might have had two stretches of sleep during the night, or else a siesta in the afternoon. The only clock they had to watch was the rising and falling of the sun.

It was the arrival of the factory working day that changed the way we sleep. Clocking on to work at a set time, staying at work for long fixed hours and working at the pace of a machine fundamentally changed the way people thought about sleep. These factories and mills were gruelling places, with fourteen-hour shifts seven days a week not uncommon, and sleep became part of the recovery process to get workers ready again for work the next day. The industrial working day created the industrial night's sleep.

Of course, working conditions have improved radically. But in many ways the productivity-obsessed open-plan office is still a white-collar version of the factory floor, only with water coolers, flip-charts and adjustable chairs. The basic principle of the working day and the recovery sleep has stayed in place. Sleep is taken as a single stretch, in a restricted period of time, to fit in with the needs of the working day.

So how might we sleep if we were left to our own devices?

Anthropologists have studied the sleeping patterns of non-industrial societies in remoter parts of Africa and have found a much more civilised approach. In such advanced communities, people sleep as unselfconsciously and for as long as they choose, without any idea of a fixed 'bedtime' or time for rising in the morning.

The members of two forager tribes, studied in Botswana and Congo in the 1970s and 1980s, were found to stay up as long as something interesting was happening and then to fall asleep just as informally when they felt tired. Sleeping was often in communal areas and people might get up again if they thought they were missing out on something. Sleep was taken as and when the urge arose, it was a relaxed occasion, with no rigid dividing line between times for sleep and waking.

How different our industrialised sleep seems. We return exhausted to our little boxes, try to make the most of a time-pressured burst of sleep and then rush back to work again. If this seems normal, then think how differently we behave on holiday, often taking long naps and switching to a much less rigid way of sleeping. It's altogether more human.

The sad thing is that we seem to be going backwards. Multinational companies setting up offices in southern Europe are more likely to impose a standardised working day, crushing the ancient pattern of the siesta. Mediterranean towns that once completely shut down in the afternoon are now peppered with shops that never have the good sense or good manners to know when to close. One of the treats of going to some dusty place in Italy or Spain used to be that everything stopped in the afternoon. There was no point trying to do anything, no point rushing, no point buzzing around like an angry tourist fly – everything was closed. It was an act of communally enforced common sense. Joining that choir of afternoon snores was as much a local delicacy as sharing the regional food.

It's not as if the urge to take an afternoon sleep during the working day is a random act of laziness. The body physically wants sleep. Mammals all over the planet are closing their eyes. There is a drop in body temperature, energy levels fall, there is a sense of fatigue, our bodies are preparing for sleep. But we cheerfully ignore the signals to slow down and rest. Like a tank column flattening the delicate ecosystem of a forest, the working day crashes on regardless, paying no heed to nature.

There are all kinds of surveys, including one, unsurprisingly, from the National Siesta Day campaign, showing how much performance at work is improved if staff are allowed to take an afternoon nap. But few employers take such an enlightened view. Something of the Victorian factory owner still lurks in the shadows, demanding a full day's work, with sleep to be restricted to the worker's unpaid hours. In the Victorian mills, children would fall asleep at their machines, which was an entirely natural reaction to being so tired. They would be beaten awake, because sleep was not allowed during the working day. We might like to think that we are much more progressive today, but if you began lying down at work and falling asleep, for how long would your bosses tolerate it?

Because work gives us money and status, it usually gets its way. We work long hours, we get home late, we start socialising too late, we go to bed too late, we don't have enough sleep, we're shaken awake by the alarm clock and the whole cycle begins again. Sleep and work should be better friends.

See also How the ancestors slept, p. 109

Einstein and the long sleepers

SLEEP IS TO BE ENJOYED, not endured; it's there to be savoured and luxuriated in, not rationed. It's one of the few things in life that has no financial cost or artificial quota. It's an infinite resource of restoration.

So whenever it's proudly declared that Margaret Thatcher only slept for five hours a night, then let's raise a glass instead to the champions of the lie-in, those noble souls who loved the pillow and loathed the ugly demands of daylight. The long sleepers need recognition too.

Albert Einstein could be the poster boy for long sleepers: individualistic, unconventional, immensely creative. Einstein was, depending on which you reports you believe, a sleeper of epic proportions, routinely sleeping for much longer than the average seven to eight hours. He claimed he needed ten hours' sleep, and eleven hours if he was working on something important.

Recognising that his mind had a habit of wandering when he was trying to concentrate on a problem, Einstein developed a napping system in which he would break off from his work, sit in a comfortable armchair and hold out a pencil in his hand. When he fell into a deep sleep, the pencil would be dropped and the sound would wake him up. Refreshed, he would return to his desk to work.

Long sleepers are usually born that way, always needing more sleep than the average. It affects about one in fifty people and it's marginally more common among men than women. Getting eight hours a day can be tough enough for most people, so finding time for ten or twelve hours' sleep every day can be a real problem, but without these extra hours the long sleeper is left feeling tired all the time. Suggestions for tackling this include taking naps during the daytime and going to bed early at night, but that might all sound a little optimistic for anyone trying to hold down a job, study for exams or bring up a family.

zzzzzzzzzzzzzzzzzz

Einstein was, depending on which you reports you believe, a sleeper of epic proportions, routinely sleeping for much longer than the average seven to eight hours. He claimed he needed ten hours' sleep, and eleven hours if he was working on something important.

There have been attempts to find personality differences between long and short sleepers. It's been said that people's sleep behaviour is as individual as their fingerprints, but some patterns are apparently shared between those who need more sleep than average, and some between those who need less. In the 1970s the psychiatrist Ernest Hartmann studied those who needed nine hours' sleep and those who only needed five and a half hours, and found some intriguing differences.

The long sleepers tended to be more creative and also to be more introverted. If they were involved in areas such as politics and ideas, they were likely to take a nonconformist, individual stance. Long sleepers were more likely to have emotional worries, to have more complicated personalities and to see themselves as rebels.

An intriguing aspect of this is that long sleepers, while asleep, appeared to move through more shifts in mood, from passive restfulness to appearing energetic and restless. They seemed to be having a more complicated emotional experience in sleep than their short-sleeping counterparts. Long sleepers also experienced twice as much of the REM sleep associated with dreaming as the short sleepers.

In contrast to the troubled geniuses of the long sleep, the short sleepers were a more contented crowd, better organised, more efficient, more likely to have conformist attitudes. These short sleepers were more likely to be busy and active, and were more comfortable at social occasions than awkward, individualistic long-sleepers. These short sleepers were often high achieving and successful.

The bad news for both these groups is that those who are substantially above and below the average in their sleep requirements are statistically more likely to die prematurely.

These psychological pen portraits of short and long sleepers are not universally accepted. It's difficult to distinguish between people who need to sleep long hours and those who just prefer it that way. Excessive sleeping can also be a temporary response to depression or illness. But it is still a thought-provoking idea that long sleepers need that extra dreaming time to work through their complicated inner lives.

See also Is too much sleep bad for you?, p. 136

These 'Sporty Boyees' were putting on their striped pyjamas in 1918.

The best-deserved sleep

IF EVER THERE was a moment when someone deserved the longest, most luxurious sleep in history, it must have been at the end of explorer Ernest Shackleton's remarkable polar rescue mission in 1916. But his reaction was even more surprising.

In a feat of stubborn survival and epic heroism, Shackleton managed to save the lives of all his crew when their ship, *Endurance,* became trapped and then was crushed by ice in the Antarctic. After living for months on shifting ice, the crew sailed in three small boats to the desolate Elephant Island. Here Shackleton left most of the crew, where they survived on a diet of penguins while waiting for his return. Meanwhile Shackleton and five others set off in a small boat to get help, performing an adjective-defying 800-mile journey across the South Atlantic, until they reached the rocky outcrop of South Georgia. This fragile 23-foot boat, patched up with parts of a sled, had sailed through mountainous seas, with its crew keeping warm inside reindeer-skin sleeping bags.

When Shackleton and his men arrived at South Georgia, they had yet another huge journey to undertake. No one had ever crossed the island before from the point where the boat had landed, but Shackleton and two of his crew set out on foot, climbing mountains and glaciers. To climb on the ice, they had removed the screws from the boat and attached them to the soles of their boots. For their diet, they survived by catching wild birds and seals.

When they eventually arrived at a remote whaling station, children ran away in fright at the explorers' appearance. They had worn the same clothes for a year, and their hair and beards had grown long and matted.

It had been almost a year and a half since they had set sail in the *Endurance*. They had survived appalling hunger and exhaustion, and slept in open boats, in caves, under an upturned lifeboat, all in freezing temperatures. Shackleton described how one crew member's sleeping bag had caught fire but the occupant's feet were so badly frost-bitten that he didn't even notice. They had also at times feared that falling asleep in such low temperatures would mean slipping into death, and had taken turns to keep watch in order to wake each other.

After all this hardship, Shackleton was given a bath, hot food and a bed for the night, sharing a room at the whaling station with his colleague Tom Crean. 'Our first night at the whaling-station was blissful. Crean and I shared a beautiful room in Mr. Sorlle's house, with electric light and two beds, warm and soft,' recalled Shackleton in his account of the journey, *South*. But then comes one of the strangest sentences you might ever find. 'We were so comfortable that we were unable to sleep.'

After all that, unable to sleep? How tired do you have to be? Months and months of continuous exhaustion, terrifying ordeals, hungry and thirsty, hiking huge distances over mountains, clawing their way over glaciers. 'We were so comfortable that we were unable to sleep.'

Over a hundred days after he had left his men, Shackleton returned to Elephant Island in a rescue ship, where all the crew were safely recovered.

See also Sleep deprivation, p. 151

Bed-in protest

IF YOU HAVE TO make a protest, then I suppose it's a good idea to be comfortable while you're doing it. In March 1969 John Lennon captured and confused world attention by staging a 'bed-in' protest for peace, during his honeymoon with Yoko Ono.

There had been plenty of sit-ins and walk-outs and bust-ups in the rebellious days of the late Sixties, but this was something different. In a hotel room in Amsterdam, the pop icon and his wife stayed in bed and talked about peace. The sheer passivity and apparent contradiction of taking action while lying down in bed seemed to both annoy and intrigue the world's media.

Lennon wasn't doing it for the money, because he had plenty. He wasn't doing it for the attention, because the Beatles star could have as much of that as he wanted. So what was he doing in bed? As he told a journalist, 'All we are saying is give peace a chance.'

This bed-in protest later moved to Montreal in Canada, where in another hotel room Lennon and dozens of supporters recorded the singalong peace anthem 'Give Peace a Chance'. In 2008 the handwritten lyrics were sold at auction for more than $800,000.

Black-and-white footage of the scenes in Montreal now look even more strange. An interviewer is invited into a madly overcrowded room to sit on the edge of the bed by John Lennon, who is smoking, wearing pyjamas and vast amounts of hair, surrounded by flowers and hand-made posters with slogans such as 'Hair peace'.

'You feel that being in bed compels more attention than if you were sitting on chairs?' Lennon is asked.

'Yes. And it makes it easier for us because we talk ten hours a day and it's more comfortable to be lying down.'

'During World War Two, what would have happened if Hitler and Churchill had got into bed?'

'I think that if Churchill and Hitler had got into bed, a lot of people would have been alive today,' said Lennon.

'Montgomery and Rommel?'

'Beautiful,' said Lennon.

See also Does 'early to bed, early to rise' really work?, p. 118

Roundheads and Cavaliers

HISTORY HAS A HABIT of going in circles. Even though the names and costumes change and the issues might sound different, subterranean divisions and fault lines can often be traced through every generation.

Roundheads and Cavaliers carried on fighting long after the Civil War finished, long after Cromwell had been buried and Charles Stuart was restored to the throne. The instincts and loyalties that divided the two sides have carried on in different guises – reformers versus absolutists, puritans versus the pleasure-seekers, romantics versus the pragmatists, industry versus ideas, provincials versus the internationalists, entrepreneurs versus artists, nonconformists versus traditionalists. These boundaries shift and it's possible to place people across both camps, which makes it all the more revealing to play which side would they have been on. Were Thatcher and Blair Roundheads or

Cavaliers? How about Nixon and Kennedy? Or the Beatles and the Rolling Stones? PCs and Apple Macs? Would today's eco-warriors be with the self-denying shaven-headed Roundheads or the long-haired free-range Cavaliers?

Sleep rests across this ancient divide, and it's no exaggeration to say that many of the contemporary attitudes and concerns over sleep can be traced directly back to the competing world views of the Roundheads and Cavaliers.

The puritanical Roundheads treated sleep with suspicion, believing it to be a form of idleness that should not be indulged, an instinct that had to be tamed, a sensuous pleasure that would only waste time. In this attitude it's possible to see the roots of the modern obsession with long working hours and the way that sleep is denied. Sleep deprivation has a long cultural tradition.

Take the example of John Wesley, the eighteenth-century leading light of Methodism, a hard-working and deeply humane man. It was said by Harold Wilson that the Labour Party owed more to Methodism than to Marxism. But what did he make of sleep?

Wesley experimented with his sleeping patterns to see whether he could reduce his hours of sleep. He managed to train himself to wake every day at six in the morning, then five o'clock and then four o'clock, until he had established this as his time for getting out of bed. In effect, he taught himself to become sleep deprived. An account written by a contemporary shows the disdain with which Wesley treated the comforts of the bed: 'He could never endure to sleep on a soft bed. Frequently at night, when he thought the bed too soft to sleep upon, he was wont to lay himself across it, and roll two or three times backward and forward, till it was sufficiently flattened and then he would get into it ... His whole conduct was at the greatest distance from softness or effeminacy.'

Needless to say, Wesley, like many of his contemporaries, had a nap in the afternoon. But he was clearly suspicious of sleep, viewing it almost as something untrustworthy, or sinful if it were enjoyed. Sleep was to be reduced to its most spartan essentials.

In the modern office, where people are either fighting to stay awake or boasting about how little sleep they need, there are echoes of this puritan wariness of giving in to sleep. If someone is thirsty, people tell them to get a glass of water, but if they're exhausted and sleep deprived, giving in to sleep would be seen as a sign of weakness. Stretching out on the desk and snoring would be seen as a moral failing.

One starchily angry puritan text written by the preacher Increase Mather captures this sense that sleeping is a guilty pleasure. The preacher, influential in New England at the end of the seventeenth century, warns that some people are deliberately going to church services in order to sleep during sermons. Accusing human nature of being 'corrupted and depraved', he says that some 'woeful creatures have been so wicked as to profess they have gone to hear sermons on purpose, that so they might sleep, finding themselves at such times much disposed that way'.

Meanwhile, the Cavaliers in their embroidered four-posters, their best wigs resting on the floor alongside empty wine bottles, held a very different view of sleeping.

There couldn't be a greater contrast to the frugal, clean-living puritans than the celebrations marking the coronation of Charles II. This royal return marked the symbolic end of Cromwell's sober republic, and Samuel

Pepys recorded the pageantry, the hard drinking and the bonfires around Westminster. At the end of the evening, his wife fell asleep sharing a bed with a female friend, while Pepys collapsed in another bed, shared with another reveller.

Pepys woke in the night to vomit. 'If ever I was foxed it was now – which I cannot say yet, because I fell asleep and sleep till morning – only, when I waked I found myself wet with my spewing. Thus did the day end, with joy everywhere.'

It is somehow appropriate that when the pleasure-loving, worldly King Charles faced a political crisis it was known as the Bedchamber Incident. The Cavalier monarch was an enthusiast for the bedroom, a louche lover of the pillow, interested in the divan right of kings as well as the divine right of kings. He had a series of mistresses, not hidden away in secret, but public figures with their own circuit of influence within the court. Such a situation appalled the king's wife, Catherine of Braganza, particularly when she was expected to accept the king's mistress as a lady of her own bedchamber. The stand-off ended with King Charles forcing his queen to accept this arrangement, but it caused damage to both his reputation and the integrity of his court. Only a few short years before, when the Roundheads had been in power, such bed-hopping would have been punishable by death; now it was institutionalised by the head of state.

Among the king's mistresses was the actress Nell Gwynn, and the type of plays enjoyed by these returned Cavaliers reflected the distance between themselves and the puritans, who had shut down the theatres as licentious and immoral. While the puritans saw the bedroom as a place of unnecessary idleness, the Cavaliers saw it as a place of entertainment. The plays in the re-opened theatres had bedroom scenes, and contained tales of rakes and fraudsters who dedicated themselves to getting into bed with the whores and heroines. These were cynical, promiscuous comedies, dark ancestors of the *Carry On* films of the 1960s.

The battle lines over the bedroom and the pleasures of sleep had been drawn.

See also Pepys's erotic dreams, p. 42

Preparing for a perfect sleep

WHAT ARE THE ingredients for a really good sleep? There are all kinds of warnings about what disrupts sleep, but perhaps we should approach the bedroom from a more optimistic direction. What are the factors that are going to increase the chances of sleep heaven?

The environment has to be right. Get rid of anything that might be a distraction in the bedroom, such as televisions, computers or mobile phones. Anything that is going to generate annoying electronic noises should be banished. It needs to be dark and soothing.

For the temperature, a survey from the hotel firm Travelodge reveals that eighteen degrees is the optimum sleeping temperature. This would be a merciful relief compared with some hotels I've stayed in, which have been hotter and drier than the Nevada desert in summer. These are usually the rooms where you find the window has been riveted shut, or that you can only open it by an inch, so you stick your nose out of this tiny gap like a dog trying to get air from a car window.

Before even entering this sleep palace, the US Department of Health suggests that the sleeper should have taken some exercise, several hours before bedtime, and that they should have had plenty of natural daylight during the day. Take a hot bath to feel even more relaxed.

Like an athlete preparing for a competition, there are warnings about diet. The National Sleep Foundation recommends eating at least three or four hours before bed-time and having only a light meal. The drink of choice is milk or herbal tea.

The NHS suggests eating a banana before bedtime, as it contains calming and sleep-inducing properties. It also suggests trying to get any worries out the way before bed-time, for example by writing down what needs to be sorted out the next day.

Other sleep-food suggestions from nutritionists include not only bananas and milk but camomile tea, honey, potatoes, flaxseeds, oatmeal, almonds, wholewheat bread and turkey.

There should be two pillows on the bed, as an extra pillow lifts the head and makes it easier to breathe and so can help anyone with constricted airways or any kind of breathing difficulty.

Travelodge also has some specific advice on the music most likely to make you fall asleep. According to a survey, Coldplay are the best band for putting people to sleep, and the ideal literary companion is David Beckham's auto-biography. I'm not convinced by this. You've painstakingly created this ideal cradle of sleep – you've exercised, taken a bath and rested in this dark, perfectly heated room, sipped milk and snacked on bananas – and the atmosphere is suddenly ruined by Coldplay and Beckham's book. The idea is to be lulled, not bored.

I'd look for a soundtrack that was both reassuring but not too overbearing, like World Service news or the shipping forecast on Radio 4. Just the words 'Cromarty, Forth, Tyne, Dogger, Fisher, German Bight' make me want to curl up and fall asleep. As for a book, I'd prefer a hugely long Russian novel that I had no chance of ever finishing. It would be one of those novels where everyone has at least three formal names plus a family name, and by the end of the first chapter you've entirely lost any idea of who the characters are. The more you struggle, the more sleep becomes irresistible.

In order to achieve an even greater sense of immersion, ear plugs or eye shades are suggested. Or there are those metal shutters that you get in some European houses that make the room utterly and completely without light. The danger with such extreme darkness is that you wake up the next morning completely disorientated and unable to see your hand in front of your face.

Tired, washed, fed and watered and lowered onto the crisp linen, in a darkened room with the shipping forecast providing its own restful end-of-the-night poetry, trying to work out who's talking to Anna Karenina, it's beginning to feel like a good night's sleep.

OK, who put the Beckham book under the pillow?

See also Sleep deprivation, p. 151

Sleeping together

THERE IS NO ESCAPING the sensual musk that hangs over the bedroom, captured in the Italian proverb that the bed is the opera of the poor. When the Hollywood censors of the 1930s tried to keep sex off the screen, they ruled that any couples getting amorous in the bedroom had to keep at least one foot on the floor. The bed wasn't just furniture, it was temptation with lace pillows. Even married couples were not allowed to be shown sharing a bed; instead they sat up in adjacent single beds, parted by the chaste contours of a bedside table.

Many of us who grew up in the 1960s and 1970s might feel the stirrings of a childhood crush when they learn that the first television show to include a couple in bed together (apart from couples who were married in real life) was *Bewitched.* In 1964 an episode showed the nose-twitchingly lovely Elizabeth Montgomery in bed with her screen husband, played by Dick York.

Movie-makers side-stepped bedroom restrictions with no lack of creativity, using visual metaphors such as scenes of trains rushing into tunnels, fireworks exploding outside windows or waves breaking onto beaches. But the need to conceal the bedroom reveals symbolic power. The place that people sleep together is the centre of their relationship; it's a source of power and intrigue. Allowing someone to share your bed is a statement of complete trust.

Going to bed with someone is one of the great delights of sleep. There is intimacy, tenderness, a baring of souls. It's where people reveal their inner selves and inner thoughts. Look at Morecambe and Wise. These comic

Breakfast in bed on an Edwardian postcard.

masters continued the tradition of pillow talk initiated by Laurel and Hardy by using the bed as a stage prop for their apparently private banter. Two men sitting up in bed in their dressing gowns thinking about the big issues in their lives, Eric puffing energetically on his pipe to show that this was nothing more than friendship. 'I always take my wife morning tea in my pyjamas. But is she grateful? No, she says she'd rather have it in a cup.'

These were the most popular comedians of their day. No one ever suggested that there was any sexual undertow to two middle-aged men having a chat in bed; it was a throwback to a time, not that long before, when friends sharing a bed or a bedroom would have been a much more common experience. In boarding houses, overcrowded family homes, army barracks and public-school dormitories, sleeping was a public rather than a private event. Now, in a culture so sharply attuned to sexuality, going to bed together is the preserve of couples.

Sharing a bed has its own hidden language. It might be romance that brings you into bed together, but after the naked passion of spring comes the buttoned pyjamas of winter. Faces look different in close-up on the pillow; you learn about the shapes people make with their sleeping bodies.

The classic sleeping position for a couple at the beginning of a relationship is to be wrapped around each other like spoons, touching in every sense of the word. There are other ways of staying close, such as the so-called 'Superman position', in which instead of wrapping both arms around the sleeping partner, one arm is pushed upwards. This has two advantages: it stops the arm falling asleep below the weight of your partner, and if things get dull it leaves a hand free to check e-mails on the BlackBerry.

A more relaxed position, but one that still maintains physical contact, is the 'leg hug', where the couple overlap legs, touching but not suffocating each other. Or else there is the Tarzan and Jane position, where the woman nestles her head into the manly chest of her mate, while he stares meaningfully at the ceiling.

According to a hotel survey, the most popular sleeping position for couples in Britain is the 'liberty pose', which is where couples sleep close together but back to back – although this sounds like a rather idealistic name for the pragmatic decision to stop clinging to each other like lovestruck limpets.

The survey also found that couples tended to stick to their own side of the bed, regardless of where they were sleeping. It's such an obviously common truth that it almost goes unnoticed how strongly we develop a sense of having a particular side of the bed. Sleeping on the other side would feel as strange as wearing your shoes on the wrong feet.

Once the first sultry wave of passion has passed, at some unspoken point, couples stop holding on to each other all night and begin to share the bed in their own favourite separate positions. This can have its own surprises. What happens if your bed partner turns out to be a quilt kicker, one of those people who gets hot even in the depths of winter and boots off the bedclothes in the middle of the night? There are also the leg danglers, who hook a limb over the side of the quilt, letting in a burst of cold air. And what are we to make of the person who sleeps in the shape of a starfish, jabbing and stretching across the divide of the bed, leaving their bedfellow clinging to the edge? Then there are those who sleep with their arms straight up, as though surrendering to an invisible enemy.

Sleeping positions are deeply ingrained – the most common being the foetal position – and once established are rarely changed. People have a strong tendency to sleep the same way each night, so once you find your partner, you have to expect them to stick to their way of sleeping for the rest of your days together.

Couples sleeping together have always had a special place in society, something protected and private. The marriage bed is literally deeply rooted in mythology. Homer tells how Odysseus made a marriage bed for Penelope which was built to include part of a living olive tree. This bed was a living creation, rooted into nature. It symbolized the way their bed and their relationship was a life force drawing strength from the earth. When Odysseus came back after twenty years of adventures, the place he and Penelope returned to was their olive-tree bed.

See also Sleep training: quack alert, p. 206

Why did even Don Juan fall asleep afterwards?

WHY DOES SEX DELIVER such a powerful urge to sleep, particularly for men? There are a number of theories. In the first place, if it's late at night there's a good chance that the lovers are already tired and they're in a big, warm, comfortable bed, so falling asleep isn't that much of a stretch. The combination of the exertion of sex and the wave of relaxation afterwards adds to the perfect recipe for sleep.

There is also thought to be a specific hormone released during orgasm that creates a profound sense of sleepiness. It's part of the post-coital recovery process. The reason that men can't leap back into action seconds after making love is because this hormone triggers a temporary shut-down. It cancels out the sexual arousal. The stronger the release of this hormone, the greater this sedating effect. It explains why men can experience such an extreme change, from the height of passion to a sleep-struck paralysis in the space of a few moments. Another theory is that some people breathe shallowly or even hold their breath during sex, and their reduced oxygen intake makes them sleepy. Or it could be the two bottles of wine you demolished over dinner.

The French talk of the orgasm as the 'little death', which is maybe a more melodramatic way of describing the post-coital paralysis. But it certainly captures the sense of toppled ardour. However, this sweetest of sleeps is nothing to be ashamed of. It has a long and noble history. Rubens' painting *Samson and Delilah* (c. 1609–10) shows a near-naked Samson slumped asleep in Delilah's lap. Botticelli's sexually charged *Venus and Mars* (c. 1485) shows Venus giving Mars an accusing look while Mars is sprawled in a state of naked exhaustion. To make the point even more subtly, a couple of little satyrs are hanging around in the background with a giant phallic lance. It even happened to the most famous of lovers, Don Juan. In Lord Byron's poem, he depicts Don Juan immersed in post-coital slumber:

And when those deep and burning moments passed,
And Juan sank to sleep within her arms,
She slept not, but all tenderly, though fast,
Sustained his head upon her bosom's charms.
And now and then her eye to heaven is cast,
And then on the pale cheek her breast now warms,
Pillowed on her o'erflowing heart, which pants
With all it granted and with all it grants.

He doesn't go on to say whether Don Juan snored.

See also Heroes under the hill, p. 232

Glorious lie-ins

THERE IS NOTHING MORE life-affirming than when the tired sleepyhead, facing the dreary prospect of getting up, is allowed to stay in bed. The lie-in is a moment of sleep salvation. It's the man facing the gallows hearing word of a reprieve. There isn't even any need to speculate on why it's such a pleasure – whether it's some subterranean foetal urge to stay cocooned, or a reluctance to leave the secluded world of inner thoughts. It is simply something to be savoured.

Nathaniel Hawthorne, the nineteenth-century writer, captured this lugubrious sense of isolation: 'You speculate on the luxury of wearing out a whole existence in bed, like an oyster in its shell, content with the sluggish ecstasy of inaction, and drowsily conscious of nothing but delicious warmth, such as you now feel again.'

Time itself seems to melt. Having missed the usual getting-up time, an hour passes in the blink of an eye. The mind is free to rove, imagining all kinds of ambitious plans. While still below the covers, all things remain possible. Novels are written, songs are strummed, perfect meals are planned, secret passions are requited, all in the moments when the day is still at arm's length. These can be the most creative moments of the day, a chance to step outside the usual hyperactive scramble where everyone is too busy to stop and think. It's in these bunkered moments in bed that you get the space to pause and take stock. You might even notice your surroundings, the way light falls through a window, the pattern of the cracks in the wall. It's a memory of childhood to see things in such detail. You might even realise how tired you are.

Whatever the dull reality that waits outside, the lie-in remains a haven of safety. Worries about work and money, fears of love and loss, are all kept at bay; nothing is more pressing than the head on the understanding hollows of the pillow. The sounds of the world might be baying at the window, police sirens and radio phone-ins might be screaming outside, but the lie-in lets you ignore all that.

This is the moment to let go. Relax and let the world rush on without you. It's a time of self-awareness, lying back and listening to your own thoughts, feeling the nuzzle of the quilt, aware that this is going to be a fleeting pleasure. Soon there is going to be a call. Until then, roll over and drink in this glimpse of freedom. It doesn't get much better than this.

See also Dreamland, p. 217

Sleeping like a statue: Florine McKinney in the 1930s Hollywood movie Night Life of the Gods.

The Nightly Journey

What happens when we fall asleep?

EACH NIGHT we go missing from ourselves, disappearing from the control of our waking mind and losing ourselves in sleep. Apart from a few chaotic, half-remembered glimpses of our dreams, we have little or no recollection of where we've been during that time.

Because we're not able to remember much about it, there's a temptation to think of sleep as a kind of void, eight hours of necessary nothingness, a restorative blank.

But that would be wrong, because sleep has a hidden life of its own, with its own elaborate patterns, its own rules and shifting moods. The body and brain undergo a series of changes. Night after night, entirely instinctively, we follow an intricate sequence of sleep stages.

This cycle begins with the first steps into sleep. Like a swimmer slowly entering the water, the sleeper is caught between two worlds. The sleeper is still awake, but the eyes are closing: sleep is approaching. This finely balanced, light drowsing might last for five or ten minutes, although time itself becomes less certain as sleep begins to take hold.

This is the transition into sleep. The brain begins to produce distinct types of slow waves as the sleeper begins to submerge. But the waking world is still in touching distance, and if someone is suddenly roused in this state they might claim not to have been asleep at all. Vivid sensations can be experienced during these moments, such as a feeling of falling or hearing someone calling. These near-sleepers might also wake up suddenly, as though jolted.

But if the sleeper is undisturbed, they move into the second stage, something that looks and feels like being asleep. The body's temperature begins to fall, the heart rate slows, and there are sudden bursts of brain activity, known as 'sleep spindles'. Like much about sleep, there is no certainty over the purpose of these sporadic flashes of brain activity. But one theory is that they could be a way of inhibiting outside distractions, locking down for a night's sleep. This second stage lasts for about twenty minutes, but it's still a relatively shallow sleep. The muscles are tense, as they would be when awake, and it's not too difficult to wake someone in this stage.

After the shallows of the first two light stages, the sleeper is now ready to push out into deeper waters. It might be half an hour or so since the sleeper first drifted off, and now, in the third stage, there begins another change, this time into deep sleep. The sleeper appears to be completely immersed: they won't hear anyone talking, and there is no apparent awareness of anything around them. This is sometimes known as 'slow-wave sleep', after the big, slow brain waves, known as delta waves, that roll rhythmically through the sleeper, and it lasts perhaps another ten minutes.

The fourth stage of sleep is when we're furthest away from our waking selves. It can take several minutes for a child in this stage to return to being awake. It's when people sleepwalk and when problems like bedwetting are most likely to happen. It's also when people talk in their sleep and suffer from night terrors.

But it's also this deepest stage of sleep, lasting anywhere between twenty and forty minutes, that is associated with restoring and healing. The body seems to need this nourishment. It's not just the quantity of sleep we need, but the quality.

If that all sounds like a day's work rather than a night's sleep, then that's only the beginning. Because having reached this complete absorption into deep sleep, the cycle then rewinds, back briefly through stage three and stage two, before reaching another and even stranger part of the night's adventure. About ninety minutes after the sleeper closes their eyes, the stage known as 'REM sleep' begins, making the first of a series of appearances during the night. If stage four is the heart of sleep, then REM is its soul.

zzzzzzzzzzzzzzzzz

In REM sleep the heart rate and blood pressure increase, breathing can become irregular and sexual arousal is common. The brain is up and running, but the big muscles of the body are at their most lifeless. It's as if the brain is switched on while the body is switched off.

The acronym 'REM' is derived from 'rapid eye movement', which characterises this stage. It is also known as 'paradoxical sleep', because the sleeper shows brain activity more in keeping with being awake. What is it that the eyes are searching for so energetically, scanning back and forth? In REM sleep the heart rate and blood pressure increase, breathing can become irregular and sexual arousal is common. The brain is up and running, but the big muscles of the body are at their most lifeless. It's as if the brain is switched on while the body is switched off.

It's also this phase of sleep that is most strongly associated with dreaming. This was discovered when researchers found that people woken at this point in their sleep cycle were most likely to be able to recall their dreams.

As well as being the seat of dreams, there are connections between REM sleep and memory and learning; this seems to be where the experiences of the waking day are processed and consolidated. Experiments on animals have shown that when they have to learn more complex tasks, their REM sleep increases as if to compensate.

Whatever goes on during REM sleep, we seem to need it. When humans are deprived of REM sleep they catch up at the first opportunity, bingeing in their next undisturbed sleep on all that REM sleep that they've missed. It seems that we crave it. Even though researchers remain unsure about its purpose, they assume that there must be a sufficiently important reason for the body to want to rectify any deficit.

For the sleeper, head still pressed to the pillow, this first burst of REM sleep might only last ten minutes. But as the night progresses, these phases of REM sleep become more sustained. The longer we stay asleep, the longer the REM stages, which could account for those delicious, rambling dreams that come with a good weekend lie-in.

But it's a remarkably busy night for someone who doesn't seem to be doing anything. After little more than a hundred minutes the slumberer will have been through seven stages of sleep, with the prospect of three or four more of these cycles before waking again. This happens each night, with variations for different age groups and individuals. It's all carried out according to unspoken instructions, a complex ritual of which we see nothing, following steps that no one has ever taught us.

The sleep cycle is a mysterious process. We should never think of it as being an empty space between going to bed and getting up again. It's a parallel life, a shadow of ourselves, a world in which we have years of experience but over which we have no authority. It's a place we go to each night but of which we remember almost nothing.

See also Circadian rhythms, p. 246

How much did Edwardian children sleep?

ARE TODAY'S CHILDREN really getting less sleep than in previous generations? In 1908 Alice Ravenhill, educationist and social reformer, carried out a study of the sleeping patterns of more than 6,000 children in English state schools. The findings are fascinating for two reasons: first, that the number of hours of sleep she believed to be needed by chidren was so high; and second, that none of these children from a century ago seemed to be getting anywhere near that amount. In fact, they seemed to be getting a similar amount of sleep to today's youngsters.

This Edwardian study assumed that a five-year-old child needed fourteen hours of sleep. For children getting up at 8 a.m., it would mean going to sleep at 6 p.m. This was the 'standard as defined by the best authorities'. For teenagers, the requirement was ten hours and forty-five minutes, which for anyone getting up at 7 a.m. would mean a strict bedtime of 8.15 p.m. But the report showed that neither the younger nor the older children were getting this amount of sleep. The younger children were only getting ten hours and forty-five minutes and the older children were only getting eight hours each night.

This lack of sleep was presented as an appalling outrage,

leaving children 'cruelly overtaxed'. The report also warned of the lack of quality of children's sleep because many were sharing three or four to a bed and had parents who only had 'defective home discipline'. Miss Ravenhill warned that the 'evil of insufficient sleep is widespread'.

Among the responses was the observation that children were now also expected to do homework, which would put further pressure on their sleeping time and could only lead to the 'disintegration of childhood'.

A century later we are still worrying about children not getting enough sleep. In the 1900s there were concerns about sleep being lost to child labour, overcrowding and poor home conditions; now we worry about distractions from the television and the internet. But the findings from the Edwardian survey might not be that different from the present state of sleep in many families.

The worry in 1908 over children's sleep in state schools had followed an earlier debate about the sleeping hours in the country's public schools. Angry letters had been exchanged in the columns of *The Times*, some arguing that boys were being sleep deprived and others that less sleep was a positive moral advantage. There were warnings from a parent who feared that even though children were going to bed at 9 p.m., they were up chatting until 10 p.m. Sound familiar?

Another letter writer claimed that lack of sleep at his public school had stunted his growth and that he had left school having reached a height of only 4 feet 10 inches. Another counterblast rejected such mollycoddling and warned against encouraging idleness. 'The boys who sleep soundest also, as a rule, work least.'

See also Shift workers and the polystyrene head, p. 181

How the ancestors slept

WHAT'S THE RIGHT TIME for going to bed? Whatever time I get under the quilt, it always feels later than I intended, the overcrowded day spilling into the night. A last look at the sour face of the alarm clock says I'm going to feel tired when it rings in the morning. Sleep always seems to be the first casualty of work.

But there's nothing inevitable or particularly natural about modern sleeping arrangements. Going to bed late in the evening and staying asleep until the morning is the pattern of the Western industrial working culture, but this single stretch of sleep hasn't always been common practice.

Before two events of dubious advantage – the Industrial Revolution and the arrival of the electric light bulb – there was nothing fixed about our sleep patterns. Instead, rather like the way we eat, the daily intake of sleep was spread out across the day. The afternoon siesta might seem to us like an exotic, sun-drenched, southern European tradition, but the habit of having more than one sleep in the course of the day was once practised right across northern Europe. The siesta is a tradition crushed by the demands of the nine-to-five industrial working day. The gentle afternoon nap is the red squirrel of sleep, hunted from its natural habitat by a pushier, more aggressive rival.

If you want to see what's natural about sleep, then look to nature. My dog Gracie, untroubled by gainful employment, falls asleep several times every afternoon, snoring like a tractor revving up in an underground car park. In that happy quivering spaniel face, we can probably see the relaxed sleep of our ancestors. This was the pre-industrial paradise for sleeping, when a comfortable afternoon slumber could be enjoyed.

According to one sleeping pattern identified by social historians, our agricultural medieval forebears would have got up at dawn, started work early, and then fallen asleep in the late afternoon. They would then have woken up in the early evening, their wits sharpened and revitalised for the big social activity of a communal meal. After eating and drinking and talking through the evening, people would have crashed out in a boozy stupor around midnight, ready to start again the next morning.

Such an afternoon sleep is simply obeying our natural instincts. We have a dip in energy in mid-afternoon and this is coupled with the sluggishness that follows a meal. Even though we're meant to be working, our bodies are yelling at us to close our eyes and rest. Perhaps unsurprisingly these views are endorsed by a research organisation called Siesta Awareness. Now wouldn't that be a place to work?

Or perhaps we should look to France, where in January 2007 the nation's health minister, Xavier Bertrand, pledged £5 million in support of better sleep, including a call for '*la sieste*' to be restored to the French working day, arguing that it was good for efficiency and safety.

There are other sleeping patterns that we've lost under the great steamroller of the standardised working day. It was once traditional to go to bed at about nine o'clock in the evening, or even earlier in winter, and to stay asleep until midnight. This was called the 'first sleep' or the 'dead sleep', when villages across Britain, without street lighting or electricity, would have slumbered through the evening in a profound darkness. After this first stretch of three or four hours, people would have awoken for a 'watching' hour or

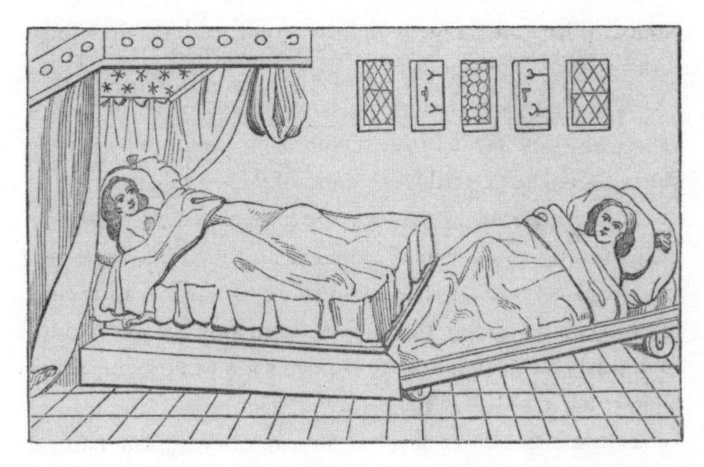

People woke for the 'watch' hour at midnight.

so. They might have stayed in bed and read or written a letter or talked. Or they might have got up and taken a snack or sat awhile and spent a quiet hour catching up on some work. This state of arousal fits in with a spike in the brain's activity at about midnight, a nocturnal burst of energy that many people experience. It was believed to be a fertile time for poets and writers to work.

Halfway between sleeping and full wakefulness, this was a special interlude in the night, a time when people felt at ease. There might be conversations, romance, prayers or contemplation. The fire might be tended or the sheer comfort of the bed might be enjoyed. After this wakefulness, people would go back to bed for the 'second sleep', which would last until daylight.

This pattern of two sleeps divided by an hour of waking might sound like a rather folksy tale. But it was widely practised in pre-industrial Britain and was documented as surviving into the nineteenth century. It's now so hard-wired into the technology-driven working week that we only sleep once a day that it's hard to imagine anything else. But the pattern of the first sleep had very long roots, mentioned in *The Canterbury Tales* and other works right through to its disappearance in the Victorian era.

This doesn't mean that it was all apple pie and thatched cottages. A study of poverty in London in 1881 recorded how the family of a self-employed carpenter was living and sleeping in one room, with six children and the two parents all taking the first sleep together in their shared bed.

But this pattern of sleeping was comprehensively trashed by the growth of factories and offices, the railways, street lights and late-night entertainment. This subtle, seductively simple way of sleeping didn't stand a chance. The employers of the Industrial Revolution wanted people to work a fixed day, with no late starts, no first and second sleeps and definitely no sleeping in the afternoon. And a single shift of sleep was what resulted.

Along with the increasing urban population and pay packets of the industrial age came the spread of gas and then electric lighting. People who had been at work during the day wanted to relax in the evening and stay up, they didn't want to go to straight to sleep. The nineteenth century saw the arrival of mass entertainment, music halls and shows that lit up the evenings. Trams and trolley-buses brought people home at hours they once would have spent deep in sleep.

If people were going to bed at midnight, there was no longer going to be an easy-going first and second sleep, no old-fashioned 'watch' hour in the middle of the night. In fact, 'midnight' was no longer the middle of the night, but only the beginning of sleep – a single stretch of sleep, sufficient to restore the workers' bodies in readiness for another day of labour.

Medical researchers in the United States have begun to pay much more attention to this lost history of sleeping patterns, not least because doctors are facing such a wave of concern from people not getting the right quantity or quality of sleep.

The growing belief is that sleeping in two shifts, this pre-industrial pattern, might be closer to the natural rhythm. The medical journal *Applied Neurology* (a light bedtime read) has examined the idea that people with sleeping disorders should try two sleeps a night, rather than take prescription drugs. For some people, the single stretch of sleep is never going to be enough, it suggests. These people go to bed too late, already sleep deprived, they fall asleep too quickly, then have an unsatisfying, restless night and then wake up exhausted.

The pattern of a first and second sleep might simply be what happens naturally when people are left undisturbed by other people's timekeeping. Experiments in the 1990s at the National Institute of Mental Health in the United States kept a group of volunteers in isolation, simulating long winter nights without electric lights, to see how people would sleep if allowed to regulate their own hours. This found that the laboratory volunteers developed a pattern of segmented sleep.

Once the structure of work is removed, watch out for the arrival of the siesta. Old people like to nap, pre-school children like to nap, holidaymakers like to nap, contestants on *Big Brother* seem to sleep half the day. It doesn't take a huge leap of imagination to realise that our ancestors, untroubled by digital timekeeping, were also likely to have enjoyed a nap.

There are plenty of advocates for the reclaiming of this traditional pattern by taking an afternoon nap. *New Scientist* published a report showing that nappers were 40 per cent less likely to die from heart disease than non-nappers. In terms of a good argument, not dying is pretty persuasive. But it's not really that simple. Even though it's a rather fashionable idea that we might all disappear for an hour to our Japanese sleep pods, it doesn't really work out like that. Sleep and work have never managed to lie comfortably together.

Sleep has always been shaped by the contemporary culture as much as by nature. The Romans liked to sleep in the afternoon and enjoyed fourteen-course evening dinners. We stay awake all day and eat sandwiches at our desk.

See also The old enemies: sleep versus work, p. 76

Saint Monday

THIS WAS ONE SAINT that factory owners and industrialists were never going to believe in. Up until the middle of the nineteenth century, many skilled workers were devoted followers of Saint Monday – the idea that Monday was a free day and not a work day. These craftsmen were used to maintaining their independence, to exercising some control over when they worked. They had never really been tamed by the fixed pattern of hours demanded by factory bosses. For these workers, Monday was a day for leisure and drinking, much to the annoyance of employers.

In the middle of the eighteenth century, more than fifty

holidays were celebrated by craftsmen – celebrated chiefly by not going to work. These traditional festivities were an overhang of medieval holy days, feast days and saints days. This unregulated way of working ignored fixed days and hours, and meant that people might sleep in the afternoon and work at night, as for example in the following description of the work habits of Birmingham craftsmen at the end of the eighteenth century: 'The industry of the people was considered extraordinary; their peculiarity of life remarkable. They lived like the inhabitants of Spain, or after the custom of Orientals. Three or four o'clock in the morning found them at work. At noon they rested; many enjoyed their siesta; others spent their time in the workshops eating and drinking.'

zzzzzzzzzzzzzzzzz

In the middle of the eighteenth century, more than fifty holidays were celebrated by craftsmen – celebrated chiefly by not going to work. These traditional festivities were an overhang of medieval holy days, feast days and saints days.

Such exotic working practices did not fit in with the need for industrialised efficiency. The idea of Saint Monday was pressed hard from two sides. There was a strain of social reformer who viewed Saint Monday as an old-fashioned, idle practice, associated with the evils of drink and debauchery. Such a reluctance to get out of bed did not sit comfortably with those who wanted to see the working class improving itself. This self-determining, unregulated way of working also faced steady opposition from the factory owners, who saw Saint Monday as an obstacle to productivity. It was a habit from pre-industrial days that went against the grain.

Although it now seems an unlikely relic, the observance of Saint Monday survived in places into the second half of the nineteenth century. But with pressure for a more disciplined workforce and the concession of a half-day holiday on Saturday, the relaxed waywardness of Saint Monday was pushed aside. Monday became the undisputed start of the working week, and Saint Monday is now all but forgotten.

See also How the ancestors slept, p. 109

Human hibernation

THERE IS NO REASON to assume that there is anything at all inevitable or natural about the pattern of modern sleeping – the idea of getting up each day at about the same time, regardless of the season or weather, in order to keep up with the demands of clocking on for a working day. Evidence suggests that seasonal sleeping was once widely practiced, with rural communities spending much more time asleep in winter or after the harvest. *The Discovery of France* (2007), a historical geography book by Graham Robb, reports how peasants in nineteenth-century France took to their beds for long stretches of the year. He quotes a civil servant in Burgundy in 1844 who found that after the wine harvest no one was doing anything that resembled work: 'These vigorous men will now spend their days in bed, packing their bodies tightly together in order to stay warm and to eat less food. They weaken themselves deliberately.'

An even more extreme version of collective human hibernation was recorded as a way of surviving the freezing winter in the Pskov area of northern Russia, where temperatures have been recorded as low as -40°C. In 1900 the *British Medical Journal* described how this communal deep sleep made it possible to eke out scarce resources:

> Not having provisions enough to carry them through the whole year, they adopt the economical expedient of spending one half of it in sleep. This custom has existed among them from time immemorial. At the first fall of snow the whole family gathers round the stove, lies down, ceases to wrestle with the problems of human existence, and quietly goes to sleep. Once a day every one wakes up to eat a piece of hard bread, of which an amount sufficient to last six months has providently been baked in the previous autumn. When the bread has been washed down with a draught of water, everyone goes to sleep again. The members of the family take it in turn to watch and keep the fire alight. After six months of this reposeful existence the family wakes up, shakes itself, goes out to see if the grass is growing, and by and by sets to work at summer tasks.

Such glimpses into the past suggest how much variety there might once have been in sleeping habits. When we want to stay in bed longer in winter, are we responding to something that would have been rooted in our ancestors' experience?

See also What is sleep for?, p. 213

Does 'early to bed, early to rise' really work?

'EARLY TO BED, early to rise, makes a man healthy, wealthy and wise.' Is it true? Roll over and go back to sleep, because this has been put to the test and found wanting. Getting out of bed early might make you feel more self-righteous, but an analysis of bedtimes and bank balances shows no connection. Those lazybones that Shakespeare called 'slug-a-beds' might be doing just as well.

But this clean-living concept of 'early to rise' certainly has long roots. The proverb was quoted in *Poor Richard's Almanack,* a collection of aphorisms and rather smug sayings compiled by Benjamin Franklin, which became a best-seller among American colonists in the eighteenth century. The proverb was also current, in almost exactly the same form, in England in the 1630s, when the self-reliant Puritans were a growing power in the land, the proverb tapping into the rather starchy belief that those who got out of bed early were morally superior and likely to be rewarded with prosperity. An even earlier version of the proverb, from the late fifteenth century, promised that early risers would be 'holy, happy and healthy'. So somewhere along the way, the ambition for holiness was swapped for wealth.

A research project at the University of Southampton in the late 1990s tried to find out whether there was any truth in the proverb by analysing the sleeping habits, wealth and health of elderly people in eight areas of Britain. It discovered a wide range of sleeping habits, with people sleeping between six hours and fourteen hours each day, and an average of about nine hours. In terms of when people took their sleep, the researchers divided people into early-rising 'larks' and late-sleeping 'owls'. The team then checked people's incomes and longevity, carried out a test of mental agility and compared the findings against sleeping patterns. The answer, in short, was that there was no connection between being an early riser and being any richer or healthier or cleverer. The proverb had the 'authority of tradition and the merit of brevity' but was completely false. The team found no patterns to suggest any advantage from being an early riser. 'If anything, owls were wealthier than larks, though there was no difference in their health or wisdom,' the researchers reported.

But such folk wisdom is hard to shake off, and the American essayist James Thurber wrote his own inversion of the proverb: 'Early to bed and early to rise makes a male healthy and wealthy and dead.' Is that any closer to the truth?

Medical researchers in the United States in 2006 tried to test these two rival sayings, pitching Franklin's early risers against Thurber's slacker's manifesto. Once again, they found no link between material well-being or improved health and getting out of bed early in the morning. 'The mortality of early-to-bed, early-to-risers did not differ significantly from other groups. There was also no relation between bed habits and local income, nor with educational attainment.'

'Early to bed, early to rise . . .' might sound like a catchy little phrase, but it's complete nonsense.

See also Counting sheep, p. 162

How much sleep do people need?

THE STANDARD ANSWER for an adult in good health is between seven and eight hours each day. But some people seem to be able to thrive on less while others need more, with surveys showing a typical spread of between six and nine hours.

It's impossible for any human to function for long without any sleep at all, but there is no fixed minimum requirement. When she was prime minister, Margaret Thatcher said she was able to survive on only four hours a night. 'You can, provided that about one day a week you do have a night when you can just have longer if you wish. But you know, it becomes so much a habit that you find you can't sleep very much longer,' she told an interviewer. About two in a hundred people are believed to be 'long sleepers', needing more than nine hours each day.

The amount of sleep we need also changes as we move through the stages of life. Babies can sleep for sixteen to eighteen hours a day, toddlers can sleep for ten to twelve hours, a young child perhaps ten hours and teenagers might need nine hours. An older person might find it harder to stay asleep at night but then might take a nap during the day.

There have been attempts to establish what is 'normal', usually by doing something distinctly abnormal such as locking people in a room and seeing how long they sleep if left undisturbed, with various permutations of light and dark. These studies come up with an average of about seven or eight hours. This might be the ideal amount, but it's much less certain how much sleep people are really getting.

There's no such thing as a 'typical' person on which to make such estimates, but there are revealing snapshots. A sample of listeners of Radio 4's *Today* programme in 2007 found an average sleep of six hours and forty-five minutes for men and seven hours and twelve minutes for women. Only 8 per cent of these morning-news junkies made the full eight hours' sleep, and both men and women were left wanting another half an hour of pillow time. Another survey of working adults in the United States, carried out in March 2008, found an average sleep duration of six hours and forty minutes. About a third of these weary workers then wanted to fall asleep in the afternoon.

z z z z z z z z z z z z z z z z z z

A sample of listeners of Radio 4's Today *programme in 2007 found an average sleep of six hours and forty-five minutes for men and seven hours and twelve minutes for women.*

Such surveys of how much time people spend asleep are usually based on some kind of individual record-keeping. But another reality-checking piece of research challenged the accuracy of such investigations. A study in Chicago in 2006 suggested that people tend to over-estimate how much time they are actually asleep.

This survey found that people were claiming to have an average of seven hours and fifty-one minutes' sleep each day. But when they were observed and monitored, they were found to be asleep for only six hours and thirteen minutes. No wonder they were feeling so shattered. Never mind missing out on eight hours, these people were not even getting seven.

The repeated conclusion from all these sleep counts is that people are getting less and less sleep, that the all-night culture is eroding the hours of rest. And there have been increasingly alarming warnings about epidemics of sleep deprivation and exhaustion. But the idea that we're all missing out on sleep and cutting corners at bedtime isn't by any means universally accepted.

There is a distinction between what people physically need in sleep and what they enjoy. Some studies have shown that people who are allowed to stay in bed longer than usual will stretch their sleeping hours to fill the available space. People like to eat when they're not hungry and drink when they're not thirsty; similarly, people often want to stay asleep longer than they might need. Just because we want to stay in bed for another half an hour, it doesn't mean that this extra sleep would be more 'natural'. Do we *need* more sleep at weekends? Or is it a case of taking advantage when the opportunity arises?

There are also contrary views on whether the average amount of sleep is really declining as much as is often claimed. Although some studies, particularly in the United States, assert confidently that the hectic, electrically powered lifestyle has reduced sleeping time by at least an hour a night, a study in Surrey of sleep habits in the twenty-first century found that the average length of sleep, just over seven hours, was the same as a similar survey had found in the 1960s. Nothing much had changed in forty years. There have also been challenges to the assumption that sleep must have been more relaxed in the past. For anyone in the nineteenth century working long factory hours and living in an overcrowded, uncomfortable tenement, sleep was no feather-bedded pleasure. Rather like life, it must have been nasty, brutish and short.

The idea of getting seven or eight hours has very deep roots. Elizabethan writer Levin Lemnius recommended eight hours; his contemporaries Andreas Laurentius and William Vaughan suggested seven. A study of working lives in the middle of the eighteenth century in England, between 1750 and 1763, shows an average sleep lasting seven hours and twenty-seven minutes. It's a remarkably consistent figure, which has survived a huge amount of social and technological change.

The answer to the question about how much sleep we need is that it varies widely between individuals. The need for sleep is said to be as individual as hair or eye colour: it's something we inherit. Sleep is 'homeostatic', which is to say it's self-regulating. When we're in need of catching up on sleep we'll fall asleep, and when we've had enough we'll wake up. It will find its own level. Sufficient for one person is red-eyed exhaustion for another. If you have enough sleep not to feel tired during the day, then that's how much you need.

See also Einstein and the long sleepers, p. 80

Caught napping

IT'S SUCH AN irresistible temptation. The eyes begin to close, there's a brief struggle with the recognition that it's only four o'clock in the afternoon, but soon the nap begins. It's a sweet stolen sleep.

Napping is also good for you. It's a rapid recharge of our jaded batteries. It can catch up on those hours of sleep missed the night before. It seems that these short bursts of sleep are a very efficient way of getting the restorative benefits.

There is plenty of serious evidence to support this. A study by Harvard scientists in 2002 found that a nap during the day provided a clear antidote to a state of exhausted 'burnout' in people who had to handle too much information and who had had too little sleep. A thirty-minute nap during a tiring day could prevent feelings of 'irritation, frustration and poorer performance' from deteriorating any further, researchers found, while a full hour's nap resulted in a positive improvement in carrying out tasks.

A separate study published in 2007, also carried out at Harvard, showed a dramatic link between regular napping and preventing heart disease. A six-year investigation found that people who napped three or more times a week had a 37 per cent lower risk of dying from heart disease.

So with all that support for napping, why is it seen as a sign of a slacker? Even the word 'nap' has a slightly cheeky feel to it, as if it is something that you get away with rather than deserve, something for a baby or a cat rather than an adult. Which is strange, because the word 'nap' comes from a perfectly respectable Old English word for sleep, which didn't have any connotation of being over in a hurry. You could nap all night before the Normans.

Even though you can find plenty of theoretical support for the benefits of napping, it's still something of a rarity in most workplaces. There might be talk about Japanese managing directors snoring at their desks and Silicon Valley blue-sky thinkers having sun loungers in their offices, but in reality it's not encouraged. The idea of sleeping during the working day goes against the corporate grain.

Maybe that's why there was a rebranding job on napping, calling it 'power napping'. It sounds a bit more macho and business school. Although of course there isn't the slightest difference between napping and power napping – any more than if you started calling porridge 'power porridge'. The principles of power napping are that time-starved, over-scheduled people can grab a short blast of sleep, restoring their energy and helping them to get through the day. Some companies, lending a little techno-credibility to all this, even have 'sleep pods' in which the power napper can lie down. These are intended to be rather hi-tech, but some look as if people are sheltering below giant tea cups. In case you mistook this for plain old catching up on a bit of sleep, these pods are billed as 'professional fatigue risk management solutions'. Yes, it's a quick kip.

zzzzzzzzzzzzzzzzz

It's better to nap for twenty to thirty minutes, maybe after lunch or in the late morning, which gives an energy boost but avoids deep sleep. When a nap spills over into a longer sleep, people feel even more exhausted when it's interrupted.

If we were free-range creatures, not battery farmed in offices, it wouldn't really matter how we napped. Our bodies would wake us when we were ready. It's an entirely natural instinct which we deny. We let our pets nap, but don't let ourselves. In the early 1800s, napping was known as a 'fox sleep', which evokes a great image of that rather deliciously sly, curled-up moment of daytime rest.

However, there are some ways in which napping can be made more effective within the constraints of the working day. Timing is an important aspect of napping. If you're at work there's no point spending too long on an afternoon nap, because there isn't time to enjoy all the courses of a full sleep cycle. Instead you need the sleeping equivalent of a snack. Too much sleep at this time will be counterproductive, leaving the napper feeling groggy.

It's better to nap for twenty to thirty minutes, maybe

after lunch or in the late morning, which gives an energy boost but avoids deep sleep. When a nap spills over into a longer sleep, people feel even more exhausted when it's interrupted. A study of nurses on night shifts has shown that the length and timing of naps is an important part of avoiding fatigue.

Another refinement is the 'caffeine nap', which claims to be a way of getting more out of a limited sleep. The napper knocks back a cup of coffee, falls asleep and then is woken soon afterwards when the caffeine kicks in. It's a kind of instant sleep with a caffeine alarm clock.

There is a long roll call of famous nappers, all proving that an afternoon sleep isn't a sign of idleness. John Kennedy, Napoleon Bonaparte, Winston Churchill and Albert Einstein were all daytime sleepers.

There is even an annual National Workplace Napping Day in the United States, which prompted one of my favourite newspaper headlines: 'Shouldn't every day be National Workplace Napping Day?'

See also Narcolepsy and microsleeps, p. 190

The temple of healing sleep

CAN SLEEP HEAL? There is plenty of evidence that not sleeping is bad for your health, but is sleep actively beneficial in curing ailments?

The ancient Greeks believed that sleep had divine healing properties, and built temples in which sick people could come to rest. Among the best known of these temples were

Sleep has been 'the great healer' for thousands of years.

the 'asclepieia', named after the Greek god of medicine, Asclepius. This particular divinity carried a physician's staff, around which was coiled a snake, a symbol of the mysterious healing powers of nature. Sick patients brought to these temples were placed in sleeping areas, sometimes with non-venomous snakes slithering around the floor. At night, if prayers for a cure were answered, divine messengers would tell the patients in their dreams how to cure their illness. The meaning might not be immediately decipherable, so priests were there to interpret the meaning of these dreams. It has also been suggested that this 'temple sleep' was a kind of trance or hypnosis, a state of deep sleep in which some medical procedures could have been carried out.

Greek gods were usually a family business, and help could also be brought to sleepers by the nursing hands of Asclepius' daughters, who included Panacea, goddess of healing, and Hygeia, goddess of cleanliness.

This was no fleeting fashion. The belief that this prayerful sleep could heal was long lasting, with the cult of Asclepius spreading to hundreds of sites across Greece and later to Rome. These were shrine centres, usually built beside a spring, to which sick people travelled long distances for a cure. It's not difficult to make a connection with much later travellers looking for healing at religious shrines or spas.

The temples themselves were often large complexes in which the priests would develop considerable expertise in using different herbs and natural medicines. This sleep therapy might seem a long way removed from the idea of modern medicine, but the original Hippocratic Oath began: 'I swear by Apollo, Asclepius, Hygieia and Panacea . . .' These ancient Greek beliefs also established the idea of the link between physical and spiritual well-being, with sleep as a kind of no-man's-land in which these worlds could meet.

See also Dream believers, p. 242

When can lone yachtswomen sleep?

A LONE YACHTSMAN or yachtswoman is a vulnerable figure, spending weeks at sea, contending with the weather, the oceans and fatigue. Sleep puts a solo sailor even more at risk, with no one to look out for danger, no one to check on the weather conditions, no one to plot the course, no one to steer. Falling asleep for eight hours isn't really an option. So the people who race yachts single-handedly have to learn how to sleep in short bursts, avoiding long stretches at night or during the day when the vessel would be unattended. It's the ultimate kind of high-stress napping: grabbing a few minutes whenever possible, constantly snacking on sleep, but never succumbing to a full night.

Ellen MacArthur, for her record-breaking, solo round-the-world trip in 2001, worked with a sleep expert on a system for taking short naps in a way that balanced her physical need for sleep against the need to be awake and alert for as much of the day as possible. The young yachtswoman slept for an average of five and a half hours each day, divided into chunks that averaged thirty-six minutes. This pattern allowed her to keep racing for ninety-four days, without succumbing to the exhaustion and confusion that comes with serious sleep deprivation.

When she talked about her sleep tactics after the race, she said that it hadn't been that difficult for her to keep waking from these short naps. 'I can't describe the mechanism that makes me wake up, I just do it,' she told reporters. 'I will sleep for forty minutes and if the wind changes I'll wake up.'

Not all solo sailors are so lucky. Others have to train themselves, using techniques such as a loud alarm that goes off every hour, forcing them to keep waking. It can take at least two miserable, sleep-disrupted weeks before the sailor is able to sustain themselves on a ration of short sleeps.

The way such solo sailors sleep has interested

researchers. This pattern of regular napping is an extreme form of 'polyphasic' sleep, in which people have many sleeps each day rather than the 'monophasic' single sleep. Researchers have wanted to examine the possibilities of surviving such prolonged sleeplessness, with a view to its application in the military or in space projects. The endurance of the long-distance sailor holds out possibilities for how the need for sleep can be tamed.

See also Sleep training: quack alert, p. 206

Hibernation

ON A COLD, dark winter's morning, when the quilt is a warm haven that's hard to leave, you have to wonder who won the evolutionary race. Bears in caves, for example, don't have to get up until spring. They might eat blackberries rather than get e-mails on BlackBerrys, they might never have watched *Newsnight*, but they're the ones looking forward to a lie-in that lasts three months.

Hibernation is a very strange process, a step beyond sleep, where life imitates death to stay alive. The body winds down to a motionless slumber, the metabolic rate falls, breathing slows, life is put on hold. Months can be spent in this condition – time passes, but the hibernating creature experiences nothing of the outside world, disappearing instead into a self-preserving internal world. Life has hit the pause button.

It really is a shutdown. The black bear can spend a hundred days like this, without eating anything, without moving, without any of the bodily functions for which bears in woods are celebrated. Think how little time we could survive without taking a drink and then consider that the hibernating bear can last all those months without so much as a drop of water passing its lips.

The reason for hibernation is to avoid the toughest, coldest, hungriest parts of the year, allowing animals to conserve their energy, stay warm and live off the body fat stored during the easier months. A form of this state of dormancy, called 'estivation', can also be observed in animals living in hot, dry regions, when they remove themselves from the arid extremes of the summer.

Although the term 'hibernation' is broadly applied to this winter shutdown, there are narrower definitions. Black bears drop their body temperature and go into a prolonged deep sleep, remaining motionless in their dens, but they could be roused. Technically speaking, they might be inactive for months but they are not 'true hibernators'. The 'true hibernators', such as ground squirrels, go even further, reaching a state perilously close to death. Their body temperature approaches zero, and their heartbeat and breathing are so slow that it looks as though they are lifeless. They can be handled without showing any sign of being alive. This is such a dangerous balance to maintain that many of these hard-core hibernators die without ever reawakening in spring.

The ground squirrel, curled in a ball in a small insulating nest, closes down everything that is not essential to maintaining this faint pulse of life. During seven or eight long months, it can lose 40 per cent of its body weight, and its bones and teeth deterioriate, as it sacrifices everything to survive in temperatutes that fall to -40°C. Most of its lifespan will be spent in this suspended state.

Living in the frozen wastelands of Alaska, the only thing that is likely to disturb the ground squirrel is a curious scientist. Ground squirrels have begun to fascinate researchers because they have found that not only does the squirrels' body temperature go down to freezing point, it can actually go below that point, to -2 or -3°C, lower than has ever been recorded for any mammal, before returning to life. To put that into context, hypothermia affects humans once body temperature falls below 35°C. The idea that animals can 'supercool' themselves and suspend almost all signs of life is being studied to see if there are any implications for human medicine.

See also Sleep and death, p. 254

How do astronauts sleep without gravity?

THE SHORT ANSWER IS BADLY. Many of them have to use sleeping tablets, because a space ship is a dreadful place to try to get any rest.

The lack of gravity means that rather than lying horizontally on a bed, astronauts often have to sleep vertically, in some kind of tethered sleeping-bag arrangement. Anyone needing the reassuring pressure of a blanket or quilt above them is going to struggle, because anything on top of the sleeper is going to float away. A pillow has to be stuck with Velcro onto the astronaut's head.

Sleeping positions are also very individual. Anyone used to curling up – and the most common sleeping position is foetal – is going to have to get used to floating loose or else having their limbs strapped into place. It's not exactly cosy.

If that wasn't enough, space travel completely disrupts the natural rhythms of day and night and sleeping and waking. The body's internal clock produces the sleep-inducing hormone melatonin as the day gets darker, which helps to lull us into drowsiness. But in space this signal is missing, as there is no natural day and night if you are hurtling round the Earth in an electrically lit tin box.

Research into the role of melatonin in regulating the body's rhythms has thrown up other questions for earth-bound non-astronauts. When people stay up late into the evening in over-lit offices or sit in front of bright computer screens at night, does that also interfere with the release of melatonin and the onset of sleep? Is this related to the health problems that can affect shift workers?

Is the struggle to go to sleep at night influenced by the confusing signals we send our own bodies?

One question – can people snore in space? – has been comprehensively answered. There had been uncertainty over this, but the microphones of Nasa scientists have firmly concluded that it is very much possible. In space, maybe no one can hear you scream, but they can definitely hear you snore.

See also Snoring, p. 200

The joy of diagonal sleeping

IT'S ONLY HONEST to admit that sleep can be a selfish pleasure. For anyone more used to sharing a bed, the occasional chance to stretch out alone, feel the wide acres of linen and experiment with elaborate sleeping positions using all four corners of this open space, is a particularly guilty delight. The usual quilt barricades have been overthrown. Tonight, it's just you and the bed spending some time together.

This was summed up magnificently in that surreal eighteenth-century masterpiece *The Life and Opinions of Tristram Shandy, Gentleman,* written by Laurence Sterne and turned into a film starring Steve Coogan. This novel pays a poignant tribute to the greatest sacrifice that accompanies marriage. You might have found your life's partner, but you have lost the delight of sleeping alone. It is the end of the private pleasure of the diagonal.

> – My brother Toby, quoth she, is going to be married to Mrs. Wadman.
> – Then he will never, quoth my father, be able to lie diagonally in his bed again as long as he lives.

Sleeping on the diagonal, apart from sounding like a dodgy art-school album title, is something that is immediately and instinctively recognisable as beneficial. You imagine the sleeper, unobstructed and indulged, flickering like a compass needle trying to find the perfect direction. The arms and legs stretch out like an opera singer, the head nuzzles into as many pillows as can be stacked beneath it: there are no boundaries for the diagonal sleeper.

See also Sleep laureates, p. 53

Does cheese give you nightmares?

THE BRITISH CHEESE BOARD – and how they must have enjoyed coming up with that name for their trade association – conducted a study to examine precisely this important question. Does cheese give you the heebie-jeebies in your sleep? Is cheese the foodstuff of nightmares?

The study, 'Cheese and Dreams', came up with the emphatic conclusion that this was a myth. A team of 200 volunteers ate cheese before they went to bed for a week, and their dreams and nightmares were monitored. There appeared to be no connection between snacking on cheese and having nightmares. Instead cheese boffins concluded, entirely impartially, that cheese contained ingredients that were conducive to a good night's sleep. Milk, after all, has been a long-standing recommendation for helping people to sleep.

The study also claimed that different types of cheese could influence sleep. Red Leicester was claimed as being particularly effective for helping sleep and creating rose-tinted nostalgic dreams. Stilton was said to be the most likely to generate bizarre dreams, and Cheshire could inspire a dreamless night's sleep.

So why do we associate cheese with nightmares? A possible connection could be Charles Dickens's nightmare-fest *A Christmas Carol,* in which Scrooge blames his strange nocturnal visions on having eaten 'a crumb of cheese' before going to bed.

However, before completely dismissing the cheese connection as an old wives' tale, there could be something behind the story. Tyramine, which acts as a brain stimulant, is found in cheese. Sensitivity or intolerance to tyramine has been linked to disturbed sleep, nightmares, headaches or insomnia. Blue cheeses are claimed to be particularly likely to cause this, giving credence to the anecdotal claims that stilton and red wine are good ingredients for a bad night's sleep. Apologies all round to the old wives.

See also Smokers' guilty dreams, p. 161

Is too much sleep bad for you?

ALL THINGS IN MODERATION. A lack of sleep is known to be a health risk, but it's also claimed that too much sleeping is linked to a shorter lifespan. A study of a million people in California found that people who slept more than eight hours a day were at greater risk of dying younger. Why this should be the case is not clear, but there doesn't appear to be any advantage to developing longer sleeping habits, beyond eight hours.

This was supported by another study carried out at the University of Warwick, which considered the sleeping patterns of more than 10,000 civil servants. This too showed the benefits of steering a middle path. Although its central finding was that too little sleep was linked to a reduced life expectancy, another less predictable outcome was that those who were sleeping in excess of eight hours were also likely to die younger.

This was something of a puzzle. A lack of sleep has been linked to a set of health problems, including obesity, high blood pressure and heart disease, but it's not clear why an extra hour in bed should increase the risk of an earlier grave.

One suggestion is that the extra sleeping time could be a reflection of other problems and that the sleep pattern is a symptom rather than a cause – that it is therefore the underlying problem that is reducing life expectancy, rather than an increase in sleep. For instance, people with serious long-term illnesses are also likely to record longer sleeping times. Another idea is that some long sleepers are in fact not getting very good quality sleep, that they have fragmented and disrupted nights, and that their long hours in bed are a sleep-hungry attempt to catch up on the rest they are missing.

In terms of how much life can be cut short by sleeping patterns, a study in Finland quantified the risks. For men, long sleepers (more than eight hours) had a 24 per cent increase in the likelihood of dying early, while short sleepers (less than seven hours) had a 26 per cent increased risk, against an average based on seven hours a day. Among women, the impact was slightly less, with a 21 per cent increased risk for short sleepers and 17 per cent for long sleepers.

But it's the kind of health information that is very hard to act upon. If you're a long sleeper needing ten hours a night, then you're hardly likely to get up two hours early, just to turn yourself into an average sleeper. Sleep is as much an individual inheritance as eye colour or the size of your feet. It's not something we can manipulate. Mind you, Napoleon was reputed to have had a no-nonsense approach to this, with the command: 'Six hours sleep for a man, seven for a woman, eight for a fool.'

See also Einstein and the long sleepers, p. 80

The bat's four-hour waking day

THERE ARE NO alarm clocks in nature. Animals in the wild follow their own rhythm of sleep and there are a remarkably diverse number of patterns. Brown bats, apart from the impressive trick of being able to sleep upside down, stay asleep for almost twenty hours each day. Looked at another way, more than four-fifths of their lifespan is spent in sleep. So which is their more significant life? Is it the inner world of those long hours asleep? Or is it the fleeting activity of four hours each day, stocking up on food for their extravagantly long sleep? In human life, we tend to think of sleep as the gap that happens between the times when we're up and about. But for a bat, it's the waking hours that are the fleeting intermission.

At the other end of the scale, bigger mammals such as elephants and giraffes get by with much less sleep. Elephants manage with only four hours a day, which they can take standing up. Giraffes, despite all that gangly effort to get around, only need a couple of hours sleep a day. Farmyard animals such as cows, sheep and horses are also relatively light sleepers, in the three to four hour range.

Cat-napping big cats like to take a longer rest. Lions, stretched out with that look of regal self-satisfaction, expect to get about twelve hours' sleep a day and with a full stomach can sleep much longer. Tigers sleep for about fifteen hours a day, cheetahs for twelve hours. Again, it raises questions about what's the 'normal' state when sleep is the majority activity during a lifetime.

In the primate world, our cousins the chimpanzees are getting about twelve hours a day, although some research has suggested that it is closer to a range of between eight and ten hours. Recordings have shown that chimps are not always asleep during their apparent bedtimes, rising several times during the night. For about twenty minutes or so they might groom themselves, eat or drink or find a more comfortable position before going back to sleep. Intriguingly, this idea of waking for a brief spell during the night was recorded as a sleeping pattern for humans in pre-industrial Britain.

Bed-making is another link with humans, with chimps making raised, bowl-shaped nests from leaves and branches. Only closely bonded chimps sleep together, such as a mother and daughter, with infant chimps sleeping in their mother's bed until about the age of five.

Giving us a strong clue about what is natural, a large majority of mammals do not take all their sleep in one stretch. They take naps, they split their sleep into chunks and they doze when they feel the need. Very few mammalian species are seen stumbling around in their pyjamas in the early hours cursing about the need to get to the airport.

There are odder aspects of animal sleep. In human sleep, the REM stage is an important and rather mysterious component, the point at which we're most likely to dream, and experience a strange unconscious arousal – humans can be as sexually aroused during REM asleep as they would be watching an erotic movie. But, who is the champion of REM sleep? It's the platypus. That strange misshapen creature with a shovel on its nose can clock up eight hours of REM sleep each day.

What is this lucky creature dreaming about? It also gets to enjoy about fourteen hours of sleep each day. How did the platypus get the winning ticket in the dream lottery? There are dozens of closely argued scientific papers examining the strangeness of the platypus's sleeping life. Not that that is likely to trouble this dream-gorged mammal.

There is not such a fortunate outcome for reptiles. While REM sleep has been detected in birds, some reptiles do not seem to have any of this dream sleep at all. Crocodiles, unsentimental creatures in every respect, have shown no sign of such dreaming. But some scientists dispute the idea that reptiles are a dream-free zone. Some lizards and iguanas, for example, are reported to display some limited REM-like brain activity when they sleep. But the question of the dreams of reptiles remains unresolved. It also leaves unanswered such cosmic Sunday afternoon questions as did dinosaurs have dreams? If so, what did they dream about?

With all questions about the nature of sleep, there's a temptation to see all other creatures through a human perspective. Sleep is well-defined in our own lives – we can see a clear separation between waking and sleeping (give or take a couple of US presidents). It's also fairly simple to detect when other mammals are conclusively asleep. But it's much less certain in other types of living things.

Insects have been studied to see if they can really be said to sleep. This also gets into complicated questions about definition. There might be patterns of quietness and withdrawal, changes in behaviour during day and night, but is that really the same thing as being asleep? Fruit flies have been poked and prodded to test their drowsiness, but there is no unambiguous answer. Even the simplest organisms follow a cycle of activity and rest, linked to the daily cycle of night and day, but is it really the same as sleeping and waking?

This uncertainty might even make us reconsider the strangeness of our own relationship with sleep. In human culture, almost everything we value and talk about belongs to the waking world, which we sharply separate from the lost hours of sleep. Sleep becomes like the pause between the words, something without a status in its own right. The other living creatures around us might experience it very differently: many of them will spend most of their lives in sleep; they might move in and out of sleep much more readily during the day than we can; the wall of separation between sleep and waking might be much more porous.

Sea creatures can spend long parts of the day in virtual immobility and then, when food or danger approach, they can move quickly before returning to a state of rest. Have they been sleeping? Or is this simply their default position, and such 'waking' moments are the aberration? We associate waking with a sense of arousal and consciousness – it's when we become like ourselves. But plants can follow the circadian rhythms of night and day, opening and closing their leaves. Does that mean that they are waking and sleeping too?

See also The old enemies: sleep versus work, p. 76

How plane noise can stress you when asleep

It's not much of a surprise that a constant disturbance from night-time noise is going to raise stress levels by keeping you awake. But research into airport noise has shown that even when people are asleep they can still show the classic symptoms of stress.

People living near to Heathrow airport and major airports in Italy, the Netherlands, Germany and Sweden were monitored through the night for increases in blood pressure. It was found that loud noises from aircraft produced an increase in blood pressure, and that the louder the noise, the greater the increase in pressure. This suggested that even though people were apparently unaware of any noise pollution, their bodies were experiencing the impact of it during sleep.

A link between raised blood pressure and noise has already been established, affecting people living near busy roads or under a flightpath. But this study has shown that for people living near airports there isn't even any respite while they sleep. It shows how much people remain sensitive to their surroundings during the night.

See also Light pollution, p. 185

The sleepless city

> For where and when in this great city, I should like to be told, can anyone secure six hours of undisturbed sleep?
>
> If insanity increases, if doctors are more busy every year with diseases of the nervous system, if men and women wear out faster and faster, who can wonder, if he will take the trouble to consider how utterly our municipal arrangements ignore the necessity for sleep?

THIS WARNING about how an overcrowded London was destroying sleep was made by a Harley Street doctor in a letter to *The Times* in 1869. We tend to think that urban worries about noise pollution and too fast a pace of life are an entirely modern experience. But this letter suggests that the battle between the need for quiet rest and the noisy restlessness of any big city has been going on for some time. It also makes a very familiar connection between such urban pressures and mental health.

What was keeping this doctor awake at night? Cabbies shouting, drunks singing songs in the middle of the street, paper boys, organ grinders, dustmen and, worst of all, concertina players. It's not exactly the chocolate-box picture of Victorian London.

See also Dreams from the 1930s, p. 239

How a dolphin doesn't drown in its sleep

IN TERMS OF evolutionary adaptations, the way that dolphins sleep must be one of the smartest. The dolphin, in common with other marine mammals, can allow half its brain to sleep, while keeping the other half awake. In this sense, the dolphin is never in a state of complete vulnerable unconsciousness. There is always at least half of the brain on lookout duty.

When dolphins enter this state of semi-sleep, perhaps for a couple of hours at a time, they float below the surface, in a condition known as 'logging' (because they look like logs lying under the water). The alert half of the brain watches out for predators and sends the dolphin to the surface for air when necessary, stopping it from drowning. The two hemispheres of the brain take it in turns to rest and stay awake, keeping one eye open and the other shut.

This isn't just a clever party trick. While breathing in humans is an involuntary response, which we can cheerfully perform while asleep, breathing for dolphins and whales is a conscious act – they need to be sufficiently awake to know when to surface and to take a breath.

There are group versions of this half-waking, half-sleeping behaviour. Ducks on the perimeter of a group resting together in a vulnerable place literally keep one eye open for danger.

Dolphins are believed to be in this restful state of semi-sleeping for about a third of each day – a similar amount of time to human sleep. And sleep researchers have speculated on what the dolphin's 'unihemispheral' sleep technique could tell us about human sleep problems. Is it possible for an animal to be sufficiently rested and refreshed without completely shutting down in full sleep? Can we get the full benefit of sleep, while remaining partially awake? Is behaviour such as sleepwalking linked to part of the brain still being awake when it should be asleep?

This half-asleep and half-awake pattern raises other questions. Can an animal have dreams while half awake? It used to be believed that dolphins did not have the REM sleep that is linked to dreaming. But more recent research has challenged this, arguing that there are signs that dolphins are able to dream. Does this mean that a drowsily awake dolphin can watch its own dreams?

See also Dreamland, p. 217

A Guinness moonlight lullaby from 1935.

Coffee versus sleep

COFFEE, or more specifically the caffeine in coffee, keeps you awake. It's a stimulant that disrupts the process of falling asleep. Most of the guidelines for getting a good night's sleep will include a warning about steering clear of coffee. But looked at another way, if you're struggling to get something done at night-time, a cup of coffee can temporarily stave off the urge to sleep.

Working out how much caffeine you're drinking isn't always that straightforward, as the levels in a cup of coffee can vary significantly. A cup of ground coffee typically contains between 80 and 115 milligrams of caffeine, while a cup of instant coffee averages 65 milligrams. Espresso might feel like it packs a punch, but because it's a much smaller volume, the caffeine dose is usually only about 80 milligrams.

But these levels fall within a very broad range. A study by the Food Standards Agency found that caffeine levels in instant coffee varied from 21 milligrams to 120 milligrams, while ground coffee ranged from 15 milligrams to 254 milligrams. This study, which tested hot drinks made at home and in cafés in ten different areas of the UK, also found that caffeine levels in cups of tea ranged from 1 milligram to 90 milligrams. Leaving the tea bag to steep for five minutes can almost double the amount of caffeine in a cup.

So in trying to gauge the impact of a drink before bedtime, you need to take many variables into account – the size of the cup, the type of coffee or tea and how strong it is. Five cups of weak instant coffee might have a lower caffeine kick than a single strong cup of ground coffee.

If you have been knocking back the coffee and it doesn't seem to stop the urge to sleep, that might be because of another aspect of the coffee versus sleep equation. People develop a tolerance to caffeine, so that the more of the stuff we knock back, the less it makes a difference. For the occasional coffee drinker, a strong slug of fresh-brewed might keep them up for hours. For the hardened coffee swigger, it's not going to have much impact.

But coffee can have a relatively long-lasting effect. Six hours after knocking back a cup, half the original amount of caffeine can still be racing round the body. So swigging a cup of coffee at teatime can still have a perceptible difference at bedtime.

It's also worth remembering that coffee isn't a substitute for sleep. Even though you can stop yourself from falling asleep, this doesn't take away any of the need. It is still there, postponed briefly, but still needing to be satisfied. Anyone who has stayed up through the night drinking too much coffee will know that awful spaced-out feeling when you're buzzing with caffeine and scratching with tiredness from lack of sleep.

The effect of coffee on sleep has been recognised for a long time. This was reflected in an espresso-sharp theatre review of a play called *Black Coffee,* written by Agatha Christie. The *Observer*'s critic, in April 1931, wrote: 'Black coffee is supposed to be a strong stimulant and powerful enemy of sleep. I found the title optimistic.'

See also: Top ten tips for a bad night's sleep, p. 163

Can sleepers commit crimes?

IN 1846 Albert Tirrell was in court in Boston in the United States facing a charge of murder. He was accused of killing a prostitute, Maria Ann Bickford, by cutting her throat, and then of setting fire to the brothel where she worked. He didn't deny that he had carried out these dreadful acts. His defence was that he had been sleepwalking during this whole incident and had no responsibility for actions over which he had no control. The jury found in his favour, acquitting Tirrell and establishing the first successful use of sleepwalking as a defence.

This trial highlights the strangeness of sleepwalking. The person is moving around and can find their way, but they are not actively conscious. They might be able to respond to a simple question, but they are not really looking at the external world and might not be able to see who or what is around them. Are they really there? How much are they liable for what they say or do?

Sleepwalking isn't the acting out of dreams, as it doesn't occur in that dream-friendly stage of the sleep cycle. Instead it happens in the deepest sleep, when the brain is furthest from its waking state and is at its least active. It's when the brain, the seat of consciousness and personality, has least control.

Usually the worst that can happen is that the sleepwalker hurts themselves or does something embarrassing. But what happens if they harm someone else? Where does responsibility fall for an act committed by someone who is unaware of their actions?

A sleep conference held by the Royal Society of

Medicine was told that violence is much more common than has been recognised in cases of 'sleep terror', a condition related to sleepwalking. Those suffering from sleep terror think they are being attacked and might kick or punch the person in bed with them. In one case, a wife had to move beds after her husband kept attacking her, believing that she was a Japanese guard in a prisoner-of-war camp.

There have been more recent cases where sleepwalking, or 'automatism', has been used as a defence. In 2005 Jules Lowe was cleared of murdering his elderly father, after the court in Manchester was told that he had been sleepwalking during the attack and had no recollection of the events. The verdict was not guilty by way of insanity.

There are believed to have been over sixty fatalities around the world in which sleepwalking has been claimed as the defence, not always successfully.

See also Dream poetry, p. 237

The meaning of 'nightmare' comes from the nocturnal horror of the incubus.

Sleep Hell

Sleep deprivation

THERE HAVE always been individuals with sleep problems, but never before has there been such widespread concern about so many people suffering from sleep deprivation. Living in a 24-hour culture is one thing, staying awake for it is another.

It doesn't feel like there are enough hours in the day to get the sleep we need. Doctors, always keen on acronyms, say that one of the most common complaints faced in the surgery is 'TATT', or 'tired all the time'. In the United States, always ahead of the curve on health worries, concern about sleep problems has spiralled. Between 2001 and 2007 there was a 75 per cent increase in the number of sleeping tablet prescriptions. In the same period, the value of sleeping-aid sales rose from $1.3 billion to $4.6 billion.

Staying late at work, bringing work home, having too many distractions in the evening – there might be plenty of reasons not to go to bed on time. But it doesn't take away the physical need for sleep. We're trying to do too much of everything except sleep.

But there is a price to pay. Missing out on sleep means neglecting the body's system for rest and repair. It makes people feel stressed and irritable, it makes concentrating more difficult, simple tasks can seem complicated, memory and motor skills are diminished. Staying awake for nineteen hours has the same negative impact on performance as being over the drink-driving limit.

Everything becomes more difficult when you haven't had enough sleep. Feelings are exaggerated, something trivial can produce an angry or tearful reaction, moods can swing up and down like a fairground dipper. You can see this most transparently in young children, whose behaviour can become horribly demanding when they miss out on sleep. Lack of sleep is like some kind of mind-altering drug for children – they turn into emotional monsters.

Tests have been carried out on behalf of the military to examine how lack of sleep affects soldiers' decision-making abilities and moral judgement. Experiments have shown that when people are very tired they make rash decisions. It's a kind of moral version of reckless driving. The US Army Institute of Research found that when people had suffered from broken sleep or no sleep, they were much more like to make 'inappropriate decisions' when confronted with sudden choices. In the context of using firearms, that could be a serious mistake.

Most of us will have been through phases when sleep was in short supply, whether it's owing to shift patterns, catching up with work, an overactive social life, revising for exams, insomnia, problems such as sleep apnea or having a baby. We might not have reached the howling-at-the-moon phase, but we all have an idea of how bleak it feels to be craving sleep, to experience that strange disembodied sense of being too exhausted to talk or think.

Here's a description of how it feels: 'In anger, I threw my pager across the on-call room, slamming it against the wall. I don't anger easily or often, but I was a paediatric resident who had been awake for thirty-six hours. The pager had gone off one time too many. Sleep deprivation had changed me from a calm, caring person into an irritable, impulsive mess.' Yes, that's the worrying part: it's a doctor writing.

zzzzzzzzzzzzzzzzzz

Experiments have shown that when people are very tired they make rash decisions. It's a kind of moral version of reckless driving.

But it can get more serious. Apart from feeling knackered and grumpy, those suffering from a lack of sleep may also find themselves with long-term health problems. The risks of high blood pressure, strokes and heart disease rise sharply with a sustained sleep deficit. Diabetes is being linked to sleep loss. Obesity is also increasingly being associated with sleep problems. Anyone who has worked night shifts will not be surprised by the physical damage caused by disrupted sleep. They will have experienced the combination of compulsive junk-food consumption, dulled reactions and a body that aches without having done any exercise.

Some extreme short-term responses can also emerge, particularly when people deliberately deny themselves sleep. Take the example of Tony Wright, from Penzance in Cornwall. In May 2007 he went for eleven days, two hours, four minutes and eight seconds without sleep. By the end of this marathon of misguided endurance, he had reported visits by 'dancing bands of giggling pixies and elves.' There is no longer any official record for such trials, because Guinness World Records fears that record attempts could damage people's health. It's reminiscent of the dance marathons staged during the Great Depression in the 1930s, when poverty-stricken couples in the United States competed for cash prizes by staying awake and dancing for as long as possible, which were banned in some states after an 87-hour sleepless dance ended in the death of a contestant.

Human rights groups have catalogued the way that torturers use sleep deprivation to disorientate and break down their victims. Prisoners will start to suffer from extreme stress, hallucinations and terrifying psychological confusion. There are no bruises or external signs of assault, but the experience for the prisoner is just as brutal. According to the Medical Foundation for the Care of Victims of Torture, such 'stress and distress' techniques can cause 'lasting, profound psychological damage'.

Stalin's political enforcers used sleep deprivation as a weapon against their prisoners. Menachem Begin, who went on to become prime minister of Israel, was imprisoned in the Soviet Union during the Second World War, and experienced this torture technique. He described the sleep-starved prisoner as follows: 'His spirit is wearied to death, his legs are unsteady, and he has one sole desire: to sleep . . . Anyone who has experienced this desire knows that not even hunger and thirst are comparable with it.'

It might not be on the same level of seriousness, but parents of young children, who have sung songs in the cold of the night to their crying children, while their head was numb with exhaustion, will feel a great deal of empathy with Begin. To be kept from sleep is to be kept from staying human.

In an article in the *Psychiatric Times*, Stanley Coren, a professor at a Canadian university, blames the invention of the electric light bulb for the reduction in average sleeping hours and spells out how prolonged severe sleep deprivation produces a 'pattern of mental deterioration that mimics psychotic symptoms'. He talks about the case of a record attempt at non-sleeping in which the person became moody, irrational and paranoid and began to think the spots on the table were insects. 'He could no longer distinguish the difference between reality and nightmare.'

Take it even a step further, and a complete lack of sleep is likely to be fatal. Experiments on rats have shown that totally depriving them of sleep leads to their deaths in about two weeks. Like much that is mysterious about sleep, why his should be isn't clear, but before the rats die they go into a complete collapse. They consume much more food than usual, they waste away, they develop lesions and ulcers on their tails and paws and become unable to regulate their body temperature.

zzzzzzzzzzzzzzzzzz

Stalin's political enforcers used sleep deprivation as a weapon against their prisoners. Menachem Begin, who went on to become prime minister of Israel, was imprisoned in the Soviet Union during the Second World War, and experienced this torture technique.

Of course, we all know the corrosiveness of missing sleep. We don't really need a checklist to tell us that it's physically damaging, psychologically stressful and literally a form of torture. But the really strange part is that we don't seem to learn any lessons from it. Sleep is always the corner that gets cut. When our workloads increase, it's sleep that loses out. Then, through exhausted eyes, we read the news on our computer screens at work the next morning about how much damage we're doing to our bodies by staying up so late each night. It's one of the few pleasures in life that costs nothing – and we still can't even keep up with sleep. We've all grown accustomed to knowing that we should get more sleep, but still we press on regardless.

Among the blizzard of recent research papers warning

about the risks of sustained sleep deprivation, Eve van Caulter, a professor from the University of Chicago, has written that 'lack of sleep disrupts every physiological function of the body'. That 'every' is pretty comprehensive. But will it mean I get to bed on time?

See also Human hibernation, p. 116

Insomnia

WHAT COULD BE CRUELLER than lying awake in bed craving sleep but being unable to go to sleep? It's exhausting, it's going to be worse in the morning, it's going to make the next day a red-eyed hell, but still sleep won't come. Daylight is beginning to creep in, sleep still hasn't arrived and you get up feeling more shattered and unhappy than when you went to bed.

Between a quarter and a third of adults are likely to suffer from insomnia at some point, with women more at risk than men and older people more at risk than younger. For many, this will only be a short-lived experience, so-called 'transient insomnia', brought on by a specific worry, which might be anything from not having the money for the mortgage to an argument at work. It can mean a few long nights staring at the ceiling, but it will pass. For some, a more sustained problem, such as worrying about an illness or a divorce, can produce 'short-term insomnia', which can last for a few months. However, the real heart-breaker is 'chronic insomnia', which means that the sufferer is unable to get any decent sleep, night after night, for more than six months. They might doze for a few hours, but there is no chance of getting the necessary quantity or quality of sleep. It's estimated that between 10 and 15 per cent of adults in Britain have suffered from chronic insomnia at some stage in their lives.

Possible causes of insomnia run to a lengthy list, which the NHS divides into five categories: physical, physiological, psychological, pharmacological and psychiatric. This highlights just how many different factors can have a detrimental impact on our sleep. A physical cause of insomnia might be backache or arthritis; a physiological cause might be your partner's snores or too much light; psychological causes could include stress about work or bereavement; psychiatric causes could be depression or dementia; and pharmacological causes could be the side-effects of medicines.

Insomnia might also be a secondary consequence of other sleep problems. Sleep apnea, for example, where someone can't breathe during sleep and keeps waking up, is a growing problem.

Lifestyle can also be a contributing factor. Working an irregular shift pattern can confuse the body's internal clock and disrupt all the usual signals for falling asleep. Not getting any exercise and then swigging coffee before bedtime is also not going to help. Although drinking alcohol can feel relaxing, it plays havoc with the quality of sleep. It's claimed that about one in ten insomnia cases are linked to alcohol or drug abuse.

Insomnia can also be a learned behaviour. It can begin with something external, such as living next door to an airport, or experiencing a crisis in your personal life, but the difficulty in falling asleep can become so regular that it is hard to re-establish a healthy sleep routine. Going to bed becomes associated with lying awake rather than with going to sleep.

There are still plenty of insomniacs who do not suffer from any of these factors but are crawling the walls each night. About a third of people with insomnia will have a history of the problem in their family, which suggests a genetic link.

But what can be done to get a better night's sleep?

Sleep experts offer consistent advice on ways to improve the chances of getting to sleep. Insomnia sufferers should set a regular pattern for going to bed and getting up, regardless of whether they feel tired. Any distractions, such as televisions or computers or unnecessary lights in the bedroom, should be taken away, to maximise the chance of a quiet, calm environment. Take some exercise during the day and avoid caffeine and alcohol for a few hours before going to bed.

For long-suffering insomniacs that will all sound infuriatingly trite, as will recommendations of relaxation therapy, assorted alternative medicines or other psychological approaches. If you've been wrestling with sleep problems for years it's unlikely to be much help to be told to go jogging and drink warm milk. Another approach is to add to the millions of prescriptions made out each year for sleeping pills. But that is only another temporary measure, as sleeping pulls can lose their effectiveness and can be addictive.

The writer F. Scott Fitzgerald, in his essay 'Sleeping and Waking', depicted the nightmarish predictability of his insomnia, describing his preparations for bed, with a drink and sleeping tablets, all the while knowing that after a couple of hours' rest he would be awake again, completely unable to force himself back to sleep. When it came to insomnia, he thought that everyone had their own personal demon. 'It appears that every man's insomnia is as different from his neighbor's as are their daytime hopes and aspirations.' The essay was published in a posthumous collection in 1945 under the appropriate title *The Crack-up*.

See also The meaning of 'nightmare', p. 217

Dormouse fat and cannabis

INSOMNIA might seem like a by-product of our anxious, restless age, but it's been around for ever. Similarly there have always been miracle cures promising to rescue the insomniac from their nightly hell.

The Greeks and Egyptians cut straight to the chase by using opium. In the Middle Ages, opium was delivered in the form of 'sleeping sponges', known by their more evocative Latin name as *spongia somnifera.* An eleventh-century recipe for this included opium, henbane, lettuce seed, mandragora (mandrake) and ivy. This potent mixture was then applied to a sponge which was held to the nostrils of the person in need of putting to sleep.

Elizabethan physicians recommended insomnia cures such as 'syrup of citrons, wormwood wine or lettuce eaten in the evening'. Or sufferers could try 'poppy, violets, roses, nutmegs, mandragora, taken in distilled water'. Poppies and mandragora also crop up as sleep drugs in Shakespeare's plays. 'Not poppy, nor mandragora, / Nor all the drowsy syrups of the world, / Shall ever medicine thee to that sweet sleep,' says Iago in *Othello*.

A more challenging Elizabethan remedy, probably borrowed from the Romans, was 'the fat of a dormouse applied to the soles of the feet'. This might sound very strange, but there is a long-established connection between dormice and sleep, right through to the snoozing dormouse in Lewis Carroll's *Alice's Adventures in Wonderland.* Dormice seem to have acquired a special association as the creatures of sleep. Another name for the dormouse is the sleep mouse, and in French the equivalent phrase for sleeping like a log is to 'sleep like a dormouse'. The word itself is associated with the French *dormer* and the Latin *dormire,* meaning to sleep. Less soothingly, the Romans used to rear dormice in special jars and then eat them as snacks when the little sleep-infused creatures were sufficiently fat.

The Elizabethan recommendation to eat lettuce as an insomnia cure has a long and persistent pedigree. From ancient Rome through to the present day, it has been claimed that lettuce has sedative properties. The fluid from lettuce stems was once know as 'lettuce opium' for its supposed sleep-inducing, painkilling qualities. Researchers have dismissed this link as unproven, but the connection seems curiously deep rooted. Recently the *Sun* newspaper recommended lettuce sandwiches as an insomnia cure.

In the 1870s, another wonder drug was advertised to stop insomnia: cannabis. A French company in London advertised its special medicinal brand of 'Indian cigarettes', sold for one shilling and ninepence a case. These cannabis cigarettes, described as 'camabis Indica', were said to 'possess remarkable virtues against asthma, neuralgia and insomnia', following recent experiments in Paris hospitals.

Even more effervescently, champagne was being advertised as an insomnia cure in the 1890s. Laurent-Perrier offered a 'Coca-Tonic-Champagne', which it promised was the 'same Grand-Vin-Brut used by the actual growers as a sympathetic vehicle for that well-known Nerve Tonic Coca. The combination is perfect and should be used in all cases of insomnia.' It assured readers that the 'medical profession strongly recommends' such wines.

Dormouse fat had by then inexplicably slipped from fashion.

See also The temple of healing sleep, p. 126

Smokers' guilty dreams

IT SEEMS LIKE a peculiarly cruel trick of the mind, but it is extremely common for anyone who has given up smoking to dream that they are back on the cigarettes. Anti-smoking advisers say that people trying to give up cigarettes can be routinely tormented by these dreams.

It's the worst of all worlds. The smoker has done the decent thing by quitting, but then is teased by their own subconscious with a vision of how much they're missing their former drug of choice. This means that the reformed smoker doesn't even have the pleasure of the sly cigarette, while still feeling guilty about their lapse. One common dream scenario is of being offered a cigarette at a party, with the ex-smoker lighting up at the first opportunity, leaving their waking self in a state of guilty envy.

These smoking dreams can go on for years after giving up and can be triggered again by stress. People giving up smoking can be taken aback by the intensity of these dreams. They can seem very realistic, the ex-smoker tasting the nicotine and smelling the cigarette smoke. Although smokers might have enjoyed cigarettes for many years and never had a dream about smoking, when they give up they don't seem to be able to stop dreaming about them. It's the nicotine's evil revenge.

See also Dream believers, p. 242

Counting sheep

COUNTING SHEEP is the classic way of making yourself fall asleep. But does it work? A study by researchers at Oxford University investigating that urgent question concluded that counting sheep is not particularly effective. They tested the idea by giving three groups of insomniacs a different sleep-inducing concept to think about. The first group had to count imaginary sheep, the second group had to think about a relaxing scene like a beach, and the third group could think about whatever they wanted.

The winner, in terms of putting people to sleep most efficiently, was the restful scene, with the sheep counting and the random thoughts trailing behind. It seems that tranquil thoughts are a more powerful deliverer of sleepiness than the sheer tedium of trying to count all those leaping lambs.

The notion of counting sheep to lull the counter to sleep is said to be derived from the difficult and repetitive task shepherds faced when making sure that all their flock was present and correct. In the north of England and southern Scotland they had their own numbering system, which was used by shepherds to make sure they had the right headcount. This was painstaking work, carried out several times during the day, with the shepherds using a rhythmic counting process to keep a tally of the sheep. It was such an endless task that it was said to send the poor shepherd into a mind-numbing slumber. This counting terminology, known as 'yan, yan, tethera' is strongly linked to Celtic languages, showing how pockets of languages can survive in isolated communities.

Shame it isn't any good for making you sleep.

See also Lullaby, p. 39

Top ten tips for a bad night's sleep

HERE ARE SOME IDEAS for making a bad night worse. Let's raise a glass to sleep hell.

Drink a load of caffeine, particularly ground coffee, just before bedtime. A few alcoholic drinks will also increase the likelihood of waking up during the night and make sure that any sleep is dissatisfying. Taking up smoking will give you the chance to add nicotine to your list of sleep-disrupting added extras.

Make sure the bedroom is either very stuffy or freezing cold – either will make it less likely that you'll be able to rest and stay asleep all night. Avoid a well-ventilated, comfortable bedroom at all costs. Wear bedclothes that are either too heavy or not warm enough, depending on the season.

Eat a vast quantity of fatty or spicy food as close as possible to bedtime. This will give your stomach something to think about. Heartburn can also be a good way of delaying sleep. Avoid anything soothing like milk.

Background noise, such as the hissing of a radio or the all-night chatter of a television, will help to disturb sleep. Lights are also useful as sleep disrupters.

Find something that is really going to stress you out and dwell on it as much as possible. See if you can get your heart racing at the same time as your mind.

Get a job that involves shift work, ideally as irregular as possible, so your body never knows if it's time to go to bed or time to get up. This can really prove a killer blow to any attempt at sleeping.

Use your bedroom for as many other distracting tasks as possible. Set up a computer beside your bed and make sure the BlackBerry is always turned on and as close to the pillow as possible, so that it never feels as though the bedroom is really a place for sleep.

Avoid any wind-down time between activities and going to bed. For instance, see if you can find some really difficult, stressful work to begin late at night, and then go straight to bed. Or start an argument with a relative. Your mind will be buzzing and sleep will be a distant memory.

Find a partner who snores like a train, then get them to have a few pints of real ale in the minutes before bed. They will sound like a plane getting ready for take-off, your nerves will be cut to shreds, and any chance of sleep will be ruined.

Exercise as little as possible and turn yourself into a run-down physical wreck. Remember: feeling depressed and a lack of well-being are among the most effective ways of promoting sleeplessness.

Enjoy.

See also Futons: beds of torture, p. 57

Mothers' ruin

WHILE YOUNG CHILDREN clutch their beakers of milk, their knackered parents clutch beakers of coffee. Sleep deprivation and parenthood go hand in hand.

It's only honest to admit that mothers often bear more of the brunt of this than fathers. While men can sleep through the sound of a baby crying in the next room, women are often much more sensitively attuned to the cries and can't stop themselves from waking up. Sometimes the baby will have barely begun to call when the mother immediately stirs.

This might not entirely be the fault of the dozing dad. There is a theory that when women become mothers they develop a greater sensitivity to noise, so that they are on standby even when their eyes are firmly shut. This is good news for the baby, but not such great news for the mother, because this sensitivity in sleep doesn't necessarily stop when the child begins to grow up.

Women are more likely to suffer from insomnia than men, and one factor affecting those who have had children could be that their sleep has been changed by motherhood:

A bedtime song for children separated from their parents in wartime.

they have become more alert to noise, and wake too easily. This might have helped them stay on duty for their baby, but years later they might not be able to sleep through the noise from traffic or the sound of a car door slamming.

There can also be tough times ahead for mothers when they go back to work. A *Mother and Baby* magazine survey found an exhausted 56 per cent of working British mothers feeling that tiredness left them in a 'state of despair'. It's a rather stark state of unhappiness – and an even higher number, 70 per cent, thought their tiredness stopped them from being able to carry out tasks effectively. By the time the baby had reached eighteen months, mothers were averaging only five hours' sleep each night. It's no surprise that they reported arguments and stress in their relationships, with lack of sleep disrupting every aspect of their lives.

Another survey compared the sleep of modern mothers with the experience of their own mothers – and claimed that mothers in the Britain of the 1960s and 1970s had 30 per cent more sleep than the current sleep-deprived generation. It's probably no surprise either that eight out of ten modern mothers in this survey were more interested in sleep than sex.

See also When can lone yachtswomen sleep?, p. 128

Dying for sleep

THE DISEASE called 'fatal familial insomnia' must be one of the cruellest illnesses ever to afflict a family. Once it appears, usually in middle age, there is only a nightmare of sleeplessness ahead. The sufferer cannot sleep, they grow more exhausted, more demoralised, they become desperately ill and then die.

As the name suggests, fatal familial insomnia is a deadly

insomnia that passes down through generations of a very small number of unfortunate families. If a parent carries this genetic burden, then their children have a one in two chance of inheriting the disease. Studies have traced the lineage of this deadly insomnia back through generations of unexplained deaths and mysterious symptoms which have affected families.

A medical account of a 53-year-old man who developed this disease shows how cruel it is to be deprived of the rest and repair of a night's sleep. This patient, a lively and sociable man, began to have more and more trouble sleeping, until he was only able to have two or three hours a night. Although this was deeply stressful, he continued to work and live an ordinary life, until the sleeplessness became even worse and he was only able to sleep for one hour each night.

The patient was trapped in sluggish exhaustion, always on the verge of a sleep that would not come. Such extreme sleep deprivation causes hallucinations and a fragmenting of personality. This appallingly sleep-starved man shivered and sweated, struggled to walk, could not carry out the simplest task, his eyesight weakened, his breathing became more laboured and his speech grew indistinct. It was the most extreme sleep deprivation imaginable, night after night, day after day. Eventually he was hospitalised, and died, unable to enjoy the restorative sleep his body and mind needed so badly.

So far there is no cure for this dreadful disease and for sufferers it is no consolation that it is such a rare condition, believed to affect only a few dozen families in the world. It is thought to be caused by a genetic disorder that can be traced back to a single individual who lived in Italy in the eighteenth century. D. T. Max, the author of *The Family that Couldn't Sleep*, calls it a kind of 'dynastic curse' which inflicts what is 'possibly the worst disease in the world'.

See also Sleep and death, p. 254

How Dickens fought insomnia with a compass

CHARLES DICKENS suffered terribly from insomnia, and one of his attempts at countering it was to sleep with the head of his bed facing due north. 'I can scarcely exaggerate what I undergo from sleeplessness,' he wrote. He also experimented with laudanum, a mixture of opium and alcohol, but although it put him to sleep it made him sick in the morning.

The idea of using a compass to sleep pointing in the right direction might seem eccentric, but it reflected the Victorians' interest in the influence of magnetic fields. And this belief that direction could influence the quality of sleep persisted long after the novelist's death. One of those compendious almanacs of household advice, called *Everything Within* and published in 1935, details the best three cures for insomnia: warm milk; a brisk walk in the fresh air; and putting the bed in a north-facing direction. 'Placing the bed so that the head is to the north is said to be a means of inducing sound sleep and this method has been tried and found successful by several sufferers,' readers were advised.

Maybe Dickens was on to something, because if you look at the feng shui advice on sleep, guess what? This Chinese philosophy of placement recommends that sleeping with the head pointing to the north is the best way of resolving insomnia.

Charles Dickens managed to put his insomnia to good purpose. Unable to sleep at night, he walked the streets of London, determined to shake off sleeplessness by his constant movement, pacing around the riverside and past deserted landmarks. His description of these night walks has a dark and almost hallucinatory quality, as the writer waits for the day to arrive. These walks lasted for hours, with a friend of Dickens recording how the writer could get out of bed at two in the morning and walk continuously until nine o'clock, and it was claimed he could cover thirty miles at a single stretch. 'I wonder that the great master who knew everything, when he called Sleep the death of each day's life, did not call Dreams the insanity of each day's sanity,' Dickens wrote in *The Uncommercial Traveller*, (1860).

Descriptions of Dickens's reading tours return again and again to the theme of his sleeplessness, which could leave him so debilitated that he couldn't move, slumped with exhaustion. The triumphs of his performances in front of an eager crowd would then be followed by his private struggle with his inability to sleep.

Dickens also took a close interest in other people's sleep. And sleep researchers have credited the Victorian novelist with providing one of the earliest accurate descriptions of sleep apnea, in the character of the perennially tired Fat Joe in *The Pickwick Papers*.

There is also a theory that the ghostly dreams of Scrooge in *A Christmas Carol* are a description of the phenomenon known as sleep paralysis, in which the sleeper feels that they are awake, but their body is still paralysed and unable to move or escape. Hallucinations and ghostly visions can accompany this state, which exists between waking and sleeping. In many cultures this unpleasant trick of the mind is associated with bad omens, witches and demons.

In a rather satisfying twist, the National Sleep Foundation in the United States has acknowledged Dickens's contribution to sleep by calling its awards for sleep research the 'Pickwick Fellowships'.

See also Dream poetry, p. 237

The insomniac prime minister who had to quit

HAROLD WILSON, Labour prime minister in the 1960s and 1970s, had no hesitation in identifying the single most important ingredient in staying at the top of the political ladder: 'I believe the greatest asset a head of state can have is the ability to get a good night's sleep.' But his predecessors were not always so lucky. And one nineteenth-century Liberal prime minister left office after less than two years, driven to the point of a breakdown by his chronic insomnia.

The fifth Earl of Rosebery might now only be an historical footnote, his brief premiership overshadowed by contemporaries such as Gladstone, Disraeli and then Lloyd George. But he had been hailed as a golden boy of late Victorian politics, well-connected, witty and charming, with a career that saw him rising effortlessly through the ranks. He had also made the sound political judgement of marrying an extremely wealthy heiress. In 1894, at the age of only forty-six, he became prime minister.

Instead of triumph, this arrival at the political summit brought something much bleaker to the surface. Rosebery suffered from terrible insomnia, perhaps caused by, or perhaps causing, a tendency to depression. Unable to rest, he was driven in his carriage round and round the streets of London at night in an attempt to make himself fall asleep.

Trying to run a country while wracked with sleepless exhaustion proved to be a waking nightmare for Rosebery. In the year he took office, his problem was already being discussed by the international press. The *New York Times* reported that Rosebery had tried to overcome his insomnia 'by the American device of drinking a glass of extremely hot water just before retiring'. Rosebery was also using something much stronger in a desperate attempt to drug himself to sleep: ever-increasing amounts of morphine. He was reported to have said that he had 'tried every opiate but the House of Lords'.

Anxious, stressed, unable to find rest, Rosebery's term of office was short-lived. Out of office, Rosebery reflected on the experience: 'I cannot forget 1895. To lie, night after night, staring wide awake, hopeless of sleep, tormented in nerves . . . is an experience which no sane man with a conscience would repeat.' He never regained his political career.

See also Bed-in protest, p. 86

zzzzzzzzzzzzzzzzz

Rosebery was also using something much stronger in a desperate attempt to drug himself to sleep: ever-increasing amounts of morphine. He was reported to have said that he had 'tried every opiate but the House of Lords'.

Sex, drugs and overdoses

THE ADVICE on sleeping pills is invariably that they are a last resort, a temporary measure that alleviates the symptoms of sleeplessness to allow someone breathing space while they tackle the underlying problems. But there must be an awful lot of last resorts out there. More than 10 million prescriptions for sleeping pills are written each year in Britain. In the United States, between 2000 and 2005, sales of sleeping pills rose by 60 per cent.

The mass-market, mass-produced sleeping pill has been around since the beginning of the twentieth century, with barbiturates remaining the classic sleeping drug from the 1900s all the way through to the 1970s. Whether it was dizzy flappers in the 1920s, suicidal blondes in the 1950s or bored Californian housewives in the 1970s, barbiturates were the sleeping pill of choice.

As a piece of trivia, the name 'barbiturates' has two completely different and contradictory claims for its origins. Barbituric acid is either supposedly named after St Barbara, on whose feast day it was discovered, or after a barmaid called Barbara, who happened to be pouring drinks when the inventor was out celebrating his discovery.

Either way, barbiturates were widely used and also widely abused. They were highly addictive and when mixed with alcohol could easily lead to overdoses. The names of this class of drugs are a darker part of the popular culture of the twentieth century. Phenobarbital, Luminal, Nembutal and Veronal are as much trademarks of an era as the names of cars. Veronal was named after the Italian city of Verona, because they were both meant to be extremely peaceful.

Nembutal has been name-checked in a Clash song. Another brand, Tuinal, was made to sound lyrical by the Pogues.

In the course of the twentieth century, more than 2,500 different types of barbiturates were developed. They were being produced by the hundreds of millions, with production peaking in the late 1940s and then again in the late 1950s. Addiction was also being mass-produced, with an estimated 135,000 barbiturate addicts in Britain in the mid-1960s, proportionately much higher than the 250,000 addicts in the United States.

Some famous early deaths were associated with barbiturates, such as those of Marilyn Monroe and Jimi Hendrix. But there were also many lesser known casualties, with deaths in Britain climbing steeply in the post-war years. In the Swinging Sixties, between 1965 and 1970, more than 12,000 people died in Britain, either accidentally or deliberately, from barbiturate overdoses.

Barbiturates began to be replaced by new types of sleep drugs which were intended to be less addictive and less likely to cause harm in an overdose. As with all medical advances, all kinds of previously undiscovered side-effects were made possible by these innovations.

Sleeping pills are no longer dished out so readily as they were in the heyday of sex, drugs and overdoses. Instead there is an emphasis on confronting the reasons for sleeplessness, such as depression or anxiety or another disorder or illness that interrupts sleep. Alternatives such as acupuncture, hypnotherapy and exercise will be recommended before the use of drugs.

The sleeping tablets that are prescribed belong collectively to a group known as 'benzodiazepines', the successors to barbiturates. These are tranquilisers designed to sedate and reduce the effect of stress, and have names such as Temazepam and Loprazolam. A more recent development has been the so-called 'Z medicines', which offer a short-term way of reducing the impact of insomnia and anxiety. Z medicines have names such as Zopiclone and Zolpidem.

As with all medicines, there are problems associated with the prolonged use of sleeping tablets. The NHS warns of difficulties with concentration, memory and attention, and advises that withdrawal symptoms can include shaking and nightmares. The internet is awash with claims and counter-claims about the benefits and drawbacks of different types of insomnia drugs. In the future, drug companies aim to produce anti-insomnia drugs that will leave the body much more quickly after inducing sleep, and that reduce the risk of a groggy sleeping pill hangover the next morning.

There is no sign that the market for sleeping pills will disappear. Even that most modern of families, the Simpsons, have used sleeping drugs. In the episode 'Crook and Ladder', when Homer tries to find a cure for his own insomnia, Bart has his own take on the end result: 'A fat, suggestible, zombie dad.'

See also Preparing for a perfect sleep, p. 91

Does booze help or hinder sleep?

ANYONE who has ever had a few glasses of wine in the evening will know how alcohol can induce a very comfortable feeling of sleepiness. It's exactly this sedative effect that makes many people who have trouble sleeping take a drink or two before bedtime. But alcohol is a bad friend for the troubled sleeper. Even though it makes it easier to fall asleep, it disrupts the quality of the sleep, causing a restless, fitful night and leaving the poor boozer waking up in the morning feeling tired. There can be a temptation to increase the dose of alcohol before bedtime, but this can create a vicious circle – with the soporific effect of the alcohol diminishing while the disruption to sleep increases. The disruptive effect of alcohol on sleep can be measured up to six hours after drinking, which means that even the swift half after work can have an impact on sleep later that night.

The boozer might not only suffer from a restless sleep – drinking before bedtime can have other, more serious and subtle, consequences. Alcohol is a big risk factor in sleep apnea – a massively growing problem as we become fatter and more sedentary as a population. Moreover, some tests have suggested that even small amounts of alcohol can disrupt the way that memories are processed during sleep. Students given puzzles to memorise and then given alcohol before they went to sleep were found to be less successful at memory tests in the morning. This could be worth remembering for any students planning to have a drink in the evening when they are revising for exams.

Something that experience might already have taught many of us, and which has been proved by researchers, is that sleeplessness can also exaggerate the effect of alcohol. If you've had a bad night's sleep and then have a drink later that day, it will be like drinking on an empty stomach.

It's even worse news for alcoholics trying to kick the bottle, because one of the most common symptoms of withdrawal is insomnia. Even after the drinker has stopped drinking, the disruption and damage to sleep remains, like some kind of clock that can't be reset. This can last for months and sometimes can never be repaired. And one of the typical reasons that recovering alcoholics slide back to drinking again is because they think at least they'll be able to get to sleep.

It's more of a nightmare than a nightcap.

See also Coffee versus sleep, p. 146

Sleep debt: a modern overdraft

The idea of 'sleep debt' is as modern as 24-hour shopping, trekking round the aisles with all the other ashen-faced nighthawks in the supermarket at midnight. It's part of the hyperactive, frantic lifestyle that we've invented for ourselves, trying to juggle work and family and never quite finding the time or energy to enjoy either. It's that mad scramble across town knowing you're already late and then falling into bed at night knowing you're already doomed to be tired in the morning.

Sleep debt is the belief that sleep loss accumulates over time and that, like a financial debt, it will have to be paid off. Going to bed too late each night builds up a debt of sleep that needs to be balanced by extra sleep at some other time. This isn't an exact science – and there are critics who say that the idea of sleep debt is unproven – but it's one more reflection of the anxiety felt over lack of sleep. What makes the idea of sleep debt different is that it suggests that there can be a cumulative cost from missing an hour or so each night. Failing to get enough sleep on a regular basis, week after week, month after month, year after year, creates a pervasive state of sleep deprivation, with consequences for physical and emotional health.

A few years ago, undergraduate students at Stanford University in the United States were given a tough message about sleep debt from experts at their own sleep disorder research centre. It claimed that tests on students had shown that 80 per cent were 'dangerously sleep deprived' and that while there was widespread awareness of the importance of physical exercise and good nutrition, there was a woeful lack of knowledge about sleep. 'If you frequently feel sleepy or drowsy in any dull or sedentary situation, you almost certainly have a very large sleep debt. A large sleep debt makes us vulnerable to apathy, inattention, and unintended sleep episodes. Errors, accidents, injuries, deaths, and catastrophes can be the result, not to mention poor grades,' students were warned by Dr William Dement. Before his retirement, Dr Dement gave his students T-shirts carrying the slogan 'Drowsiness is red alert'.

The mental health charity Mind has also pointed to the corrosive effect of sleep debt on mental health: 'Sleep debt can affect intelligence and control of movement, and can have a bad effect on the metabolism and on hormones. While people are in the sleep-debt state, they are more likely to make mistakes or act irrationally. Sometimes, lack of sleep contributes significantly to the development of serious mental health problems.'

Now that the idea of sleep debt has been introduced, staring balefully at us like some kind of sleep mortgage, the next question is, How do we pay it off? Is it an hour's sleep for an hour lost, or is there some kind of interest rate? Is it like borrowing on credit cards that become harder and harder to clear?

It's not surprising that people will sleep longer, given the opportunity, after they have missed sleep in the previous nights. But there is no clear consensus on payback time for extended sleeplessness. There are claims that three days of regular sleep is insufficient compensation for a sustained loss of sleep. But another view is that sleep loss is more like a thirst, that once you've had enough to drink, the need is removed. There's no need to keep drinking more after the thirst has been quenched.

There's also another theory that the whole idea of sleep debt is an artificial invention, a product of the chattering classes' need to worry about something. The sleep-debt sceptics argue that humans are extremely adaptable in their sleeping patterns and that it's a fashionable exaggeration to imagine that we're a generation starved of sleep. This viewpoint also highlights an important ambiguity about sleep: just because people will lie in bed when given the chance, it doesn't necessarily mean that they are exhibiting a biological need for more sleep. People might sleep longer because they like sleeping, or because they don't want to get up and do something else – it doesn't mean that a longer sleep is more 'natural'. The amount of sleep we have can be compared to the amount of food we consume: both owe as much to pleasure as they do to necessity. But how can we tell the difference between sleep debt and feeling fashionably tired?

There is a widely used screening test for measuring levels of sleepiness, known as the Epworth Sleepiness Scale, named after the hospital in Melbourne, Australia, where it was developed. This is designed to distinguish those with genuine sleep problems from those with more low-level tiredness. The test gives a list of eight day time activities and asks people how likely it is that they will doze off in these situations. A four-stage scale is marked against each situation:

No chance of dozing = 0
Slight chance of dozing = 1
Moderate chance of dozing = 2
High chance of dozing = 3

The eight situations are:

Sitting and reading
Watching television
Sitting inactive in a public place (such as a theatre or meeting)
As a passenger in a car for an hour without a break
Lying down to rest in the afternoon when circumstances permit
Sitting and talking to someone
Sitting quietly after a lunch without alcohol
In a car, while stopped for a few minutes in traffic

A total score of 0 to 6 means that you have sufficient sleep, a score of 7 to 9 is average, and above 9 means that there could be a problem with lack of sleep.

See also The old enemies: sleep versus work, p. 76

Shift workers and the polystyrene head

IT'S SOMETIME in the middle of the night, there's a fluorescent light flickering above your head, the computers are drowsing on their screensavers and the photocopier is on its sleep settings. But you're awake on a night shift, trying to stay alert and purposeful when your body is screaming at you to lie down, get some sleep and behave like any other normal creature on the planet.

After a phase of caffeine-alleviated exhaustion, you start to binge eat. You're not hungry, in fact your stomach is as confused as your brain. But you eat something fat and unhealthy and probably wash it down with another coffee, then try to wake up a little, maybe chat to someone or read something distracting. It's even weirder when you're on a break during a night shift, because what are you going to do? Look around the shops, go for a run? You sit in a chair, behind yet another coffee, looking like you've been lobotomised, watching the clock stretch out the time.

Then, at about six in the morning the first pale signs of life start up again in the outside world: buses begin to trundle past, daylight appears and the new day finally cranks into action.

It's about this time that the polystyrene head begins to take hold. It's a peculiar numbness, a lack of feeling coupled with a terrible sense of vulnerability. You could easily bump into things or knock stuff over, traffic is harder to negotiate and everything seems too bright. You feel light-headed but incredibly heavy at the same time. Simple tasks become harder and harder – you have to consciously make yourself concentrate and focus on whatever job you're

Sleepless cities: Telegraph clerks processing messages in 1900.

meant to be doing. You're exhausted but also buzzing with an artificial, edgy energy.

Then, as everyone else begins to arrive for the day, it's time for you to go home, catching the train out when everyone else is coming in. It's meant to be your bedtime now, even though you won't necessarily find it possible to sleep. Being too tired to do anything isn't the same thing as being able to go to sleep. It might be nine o'clock in the morning when you get home, full of coffee and junk food, not really able to eat a proper meal and finding it difficult to relax. Who wants to put themselves to bed immediately after work in the evening? It's not exactly time to have a glass of wine and wind down. Flick on the television and the children's programmes are just starting. But if you don't get into bed in the morning, then by the time you do go to bed and get up it will be evening and time for work again, for another night in the polystyrene head.

In an all-night, always-open culture more people than ever are having to work these night-owl hours, whether it's in retail, call centres, media, transport, hospitals or the emergency services. But there is a very serious price to pay for ignoring the natural urge to sleep. The starkest and least ambiguous warning is from the World Health Organisation, which at the end of 2007 identified shift working that disrupted sleeping and waking patterns as 'probably carcinogenic to humans'. Its study highlighted concerns about a heightened risk of breast cancer among long-term night workers.

The International Agency for Research on Cancer, the cancer research agency of WHO, explained its findings:

> The studies are consistent with animal studies that demonstrate that constant light, dim light at night, or simulated chronic jet lag can substantially increase tumour development. Other experimental studies show that reducing melatonin levels at night increases the incidence or growth of tumours.
>
> These results may be explained by the disruption of the circadian system that is caused by exposure to light at night. This can alter sleep-activity patterns, suppress melatonin production, and disregulate genes involved in tumour development. Among the many different patterns of shiftwork, those that include nightwork are most disruptive to the circadian system.

This international health body revealed that nearly 20 per cent of the workforce in industrialised countries now had to undertake shift work of some kind and that there was a need for further study to examine cancer risks linked to disrupted sleep patterns.

The fact that shift patterns in many jobs are irregular provides an extra slice of sleep hell. Night shifts can be interspersed with day shifts in a way that leaves the human body completely unable to keep up with the disorientating changes involved. The circadian rhythms of sleeping and waking can take between seven and fourteen days to become established, and during that time sleep will be disrupted and fragmented, leaving the worker feeling groggy and irritable. If the shift pattern changes again soon afterwards, the whole upheaval of re-establishing a sleeping routine will also begin again. Such rotating shifts are the most dizzying and disruptive for sleep. We also know they're the most dangerous.

A study of different shift patterns on North Sea oil rigs found that those working a split-shift system of seven night shifts and then seven day shifts had an increased risk of heart disease and diabetes. A Dutch study also found night workers more susceptible to developing an irregular heartbeat.

What's most alarming about this is that even though working night shifts is clearly a potentially lethal health risk, many people are still expected to do so. Should employers require people to work in a way that knowingly increases their risk of cancer? As consumers, should we expect other people to stay up all night so that we can talk to a call centre about our overdrafts at four in the morning?

See also Circadian rhythms, p. 246

Light pollution

It's appropriate that the inventor of the light bulb, Thomas Edison, had a reputation for hating sleep. The spread of electric light has changed houses, streets and cities for ever and turned the night into a place where complete darkness is a rarity.

The darkness of night is the natural environment of human sleep. But like other natural habitats, the night sky is under threat. A constant electric glow hangs above cities. A study of light pollution published in 2007 found that skywatchers in Britain could see only a shrinking number of stars in the constellation of Orion. This is one of the great landmarks in the night sky, and for the first time in human history it is disappearing from view. We're now more likely to see the sign of Ryanair crossing the night sky than a sign of Orion.

At night-time the skies are full of light. In cities some offices never seem to turn off their lights, and these, together with advertising hoardings, street lamps, motorway networks and all-night traffic, all throw up a smog of light. According to the Environment Agency, England is one of the most light-polluted places in Europe. This has unexpected consequences for living things, which often depend on the cycle of light and dark as a trigger for their own internal clocks. The 'false dawn' of electric light disrupts bird behaviour, and trees and plants are thrown out of their natural rhythm for flowering and losing leaves.

Observatories now have to be built as far away as possible from urban sprawl so that astronomers have a chance of seeing the stars. One observatory has been built in the place claimed to have the least light-polluted skies in England: Kielder, in Northumberland.

But what does all this do to a good night's sleep?

The sensation of becoming sleepy at night is linked to melatonin, a hormone released when it becomes dark. Melatonin levels remain high during the night, before dropping again when daylight breaks. When there is no darkness, melatonin release is suppressed and this cycle of entering sleep is disrupted. It's not only outside the house that light is seeping everywhere. Inside our homes there are more lights and electric gadgets throwing out light than ever, some of which are hardly ever switched off. There is also the intrusion of light from outside, from street lighting and passing traffic. Complete darkness can be as rare within the home as it is outside.

Without this darkness, the sleep-inducing melatonin is less likely to be released. When we interrupt the cycle of light and dark we interrupt the internal rhythm of sleeping and waking. Sleep needs to be kept in the dark.

See also How the ancestors slept, p. 109

Heart of the night

THE SLEEPING HABITS of more than 10,000 civil servants have provided some of the most direct evidence of the link between healthy sleeping and healthy living. The study of this group of Whitehall's finest has shown that a lack of sleep can double the risk of dying from heart disease.

It might be self-evident that not sleeping makes you tired and grumpy, and it might be a predictable outcome that regularly missing sleep isn't good for you. But recent research is increasingly looking in much finer detail at the consequences. It's a little like going from the stage of recognising that smoking is a health risk to itemising the precise damage.

zzzzzzzzzzzzzzzzz

When people get home later in the evening, they still want time to relax and socialise. However, what the results of this study suggest is that stealing a couple of hours from sleep too often can have lethal consequences.

The study by researchers from the University of Warwick and University College London used sleep records kept by civil servants in the mid-1980s through to the early 1990s to examine how sleeping habits affected life expectancy, based on what had happened to these civil servants by 2004. After other health and age factors were taken into account, it found that those with reduced sleeping hours of five hours or less were 1.7 times more likely to have died. In particular, they were twice as likely to have died from cardiovascular problems.

This has far-reaching implications. Should lack of sleep be taken much more seriously as a life-threatening health matter? Do we neglect sleep too easily and assume there are no long-term consequences?

It's not difficult to see how sleep gets reduced. It might be long hours at work, a long commute or an attempt to catch up with an overcrowded family life. When people get home later in the evening, they still want time to relax and socialise. However, what the results of this study suggest is that stealing a couple of hours from sleep too often can have lethal consequences. 'Fewer hours sleep and greater levels of sleep disturbance have become widespread in industrialised societies. This change, largely the result of sleep curtailment to create more time for leisure and shift-work, has meant that reports of fatigue, tiredness and excessive daytime sleepiness are more common than a few decades ago,' Professor Francesco Cappuccio from the University of Warwick's Medical School told the British Sleep Society.

Sleep might cost nothing and feel like a movable feast, but Professor Cappuccio's research suggests that 'lack of sleep has far-reaching effects'.

See also: How plane noise can stress you
when asleep, p. 141

Alternative insomnia therapies

BEFORE REACHING for the pharmaceuticals, you could try one of the many other approaches to reducing insomnia. Anything that makes sleepers more relaxed, less stressed and improves their breathing or fitness is likely to help reduce insomnia. It might not resolve the underlying problems, but it might lessen the symptoms. And there are plenty of therapies and tactics that claim to help.

Yoga can improve relaxation, breathing and posture and promote calm, all likely to aid sleep. The martial art of tai chi also provides exercise and emphasises calm, controlled breathing. Simple breathing exercises and meditation are said to be beneficial too.

Aromatherapy also has adherents. The idea is that aromas from essential oils can trigger a natural healing process, evoking different thoughts and memories, which affect mood and the sense of well-being. For insomnia, lavender, rose, cypress, marjoram, camomile and sandalwood might be used.

Herbal remedies are also on offer to soothe the sleepless, which might include St John's Wort, poppy, camomile, lemon balm, hops and valerian. Ginseng tea was being promoted as an insomnia cure as far back as the 1790s.

Acupuncture provides another type of therapy. This is based on the philosphy of natural balances within the body; a problem such as insomnia therefore reflects an imbalance. Treatment is by the insertion of very fine needles into particular points in the body, with the aim of restoring balance and the flow of energy.

Hypnosis and self-hypnosis also offer help with insomnia. Rather like helping people to give up smoking, it's about getting people to change their habits and to alter their patterns of thinking and behaviour. If trying to go to sleep is associated with a stressful, negative experience, hypnotherapy might try to create a more relaxed mood and a more positive frame of mind about sleep.

Reflexology, which uses a kind of foot massage to resolve problems in other parts of the body, claims to be of benefit to insomnia sufferers. Body massages have also been used as a way of encouraging a more sleep-friendly state of relaxation. Osteopathy can provide another approach, addressing back and muscle problems that might be interrupting sleep.

Of course the veteran insomniac, having tried all these and failed, might now be sat scowling at all these claims. As worries about insomnia have grown, so has the market selling promises of sleep.

See also Hot in bed: the electric blanket, p. 70

Narcolepsy and microsleeps

Narcolepsy is a curious ailment in which the sufferer suddenly falls asleep without any warning. If you've ever met anyone with narcolepsy, which affects about one in 2,000 people, you will have seen how rapidly and uncontrollably this can happen. It can be mid-sentence when the head starts nodding and the eyes suddenly close and the sufferer dips into a narcoleptic nap. There is no way for the narcoleptic to fight this sudden onset of sleep.

There isn't any cure for this condition, although sufferers might be encouraged to take a number of planned naps during the day, as a precaution against involuntary sleeps. It's also not something that is identifiable from birth, with most narcoleptics diagnosed between the ages of fifteen and twenty. It's also possible that this sleep disorder can develop in middle age.

It can be a debilitating problem for sufferers, making it difficult to keep jobs or carry out everyday tasks such as driving a car. It must also be tough having an illness that other people see as an intriguing oddity. A few years ago there was a heart-wrenching documentary about the struggles of narcoleptics, which included their epic attempts to hold a meeting in which people kept falling suddenly and spectacularly asleep. Although this was a brave attempt at opening up a private world to the public, a clip ended up being included in a greatest television moments show and was met with laughter by the studio audience. Apologies had to be made all round. It's difficult to think of any other incurable illness that could have been held up as creating a classic comedy moment.

A microsleep is another kind of involuntary sleep. It occurs when someone is seriously sleep deprived and their body has decided that it isn't going to wait. This instant burst of sleep can last from a fraction of a second to a few minutes, and the person experiencing it might not even know that it's happened. This can be extremely dangerous if it happens to someone driving a car – those making long, boring journeys are particularly prone to being sent into a half-awake trance, which for a few seconds can dip into sleep. These moments can be enough to cause an accident.

Microsleeps can occur whenever someone is desperately tired, but there are two peak times for this to happen: mid-afternoon and just before dawn. This pre-dawn moment, when roads are empty and the day has still to wake up, is identified as a particularly vulnerable time for sleep-starved drivers.

Car manufacturers have been working on plans to detect such microsleeping. Toyota announced plans for a 'pre-crash safety system' which would include a camera behind the steering wheel that would monitor the eyes for any sign of closing. Mercedes has been working on a system for identifying changes in driving habits that could be a sign of microsleeps and fatigue, making the sobering point that during these microsleeps the car is effectively hurtling along at speed without a driver.

This type of instant napping doesn't only happen to drivers, however it can happen during the day at work, too. Someone hunched over a computer screen might have a series of microsleeps, their head nodding and glazed eyes dozing in short bursts. They might wake up with a jolt, unsure of what's happened, before returning to their task, not realising that they have briefly fallen asleep.

When people are tired enough, this micro-napping is an automatic response. Experiments on people who have been deliberately sleep deprived show that they can't be stopped or stop themselves from stealing back moments of this lost sleep. They will slip suddenly into this unresponsive sleep state.

The restorative process of sleep cannot be held back. If it isn't allowed in conventional sleep, then it will force its way into wakefulness. The microsleep is a strange and uncertain borderline where sleeping and waking overlap.

See also How a dolphin doesn't drown in its sleep, p. 143

Fat chance of a rest

ALTHOUGH SLEEP deprivation and obesity might seem like odd bedfellows, researchers have begun to make connections between these two modern epidemics. A series of studies have found a link between a lack of adequate sleep and a higher risk of obesity. But why should this be the case? Why should staying up too late or getting up too early be associated with weight gain? If anything you might assume that staying out of bed would use up more calories than sleeping.

A comparison of children in ten cities in the United States in 2007 found that 12 per cent of eight-year-olds who slept for ten to twelve hours were obese, but for those who only slept for nine hours the obesity rate was 22 per cent. The longer sleepers were more likely to be a healthy weight.

This pattern was reflected in a long-term study in Ohio

which tracked 70,000 women over a period of sixteen years. Women who slept less than five hours a night were already likely to be heavier at the start of the research period, and over the years they were more likely to put on weight than women who slept for a longer time. The sleep-deprived, five-hour sleepers were a third more likely to put on an extra 15 kg over that time, which is more than two stone.

z z z z z z z z z z z z z z z z z

Why should staying up too late or getting up too early be associated with weight gain? If anything you might assume that staying out of bed would use up more calories than sleeping.

Another research project, run by the US National Institute of Mental Health, looked at a group of 500 adults over thirteen years and found that those who slept the least were most likely to put on weight.

There are a couple of theories about this. It has been suggested that a lack of sleep could be a counterpart of a lack of exercise, reflecting on a generally unhealthy lifestyle. Youngsters who never get to bed might be up all night playing computer games and might be too tired for physical exercise. Another theory is that a lack of sleep creates hormonal changes which affect the appetite. The body's internal signal for feeling full up is disrupted. This means that sleepless people can have a kind of false hunger, continuing to eat when they have no need for food.

The whole slimming industry might be rather uncomfortable with the idea that getting a couple of hours' extra sleep is a cost-free way of avoiding looking like a midnight choc-chip muffin.

See also Hibernation, p. 130

Sleep apnea

TWENTY YEARS is a long time – imagine all those years without a single good night's sleep. No rest, night after night, month after month, year after year, decade after decade . . . it's almost too exhausting to think about.

Philip Skeates from Swindon had averaged only fourteen minutes a night for all those long years before he was diagnosed with sleep apnea. Tests showed that he was woken by this sleeping disorder, a condition that causes the airways to close, ninety times an hour – another eye-watering statistic. In a masterpiece of understatement he described his state of mind as follows: 'I was really quite grumpy.'

After the diagnosis of sleep apnea he was able to use a breathing mask that allowed him to experience seven hours of sleep for the first time since he was a teenager. These masks, attached to an air-pumping machine, will undoubtedly become a much more familiar sight as sleep apnea is more widely recognised. It is currently thought to affect about 3.5 per cent of men and 1.5 per cent of women.

'Apnea' comes from the Greek for 'without breath' and it is a condition that causes a series of short stops in breathing during sleep, often dozens of times an hour, because of a temporary obstruction in the airways. The most common symptom of sleep apnea is loud snoring, but in worse cases it can also be choking and gasping as the sleeper tries to catch their breath. The sufferer will have restless, unsatisfying sleeps and although they may not remember their night-time troubles, they can wake feeling exhausted and unrefreshed. Excessive tiredness during the day can be as much a symptom as anything that happens during the night.

The classic sleep apnea sufferer is middle-aged and overweight, and men are more likely to be sufferers than women. It's a fast-growing condition, one of those ailments that reflects the changing shape of the population. In the United States, the market in equipment and treatment for sleep apnea is expected to double to $4 billion in the next four years.

If left untreated, it can be a serious problem. Apart from the disruption to sleep, the constant stop-starting of breathing is not good for blood pressure and the heart. Moderate to severe apnea sufferers are three to four times more likely to have strokes than those without apnea. Even more depressingly, stroke victims who have sleep apnea are likely to die sooner than those without apnea. And the bad news doesn't end there.

Having sleep apnea raises the risk of having a heart attack by 30 per cent, and there is also a higher risk of being involved in a severe car crash.

Recognising the problem usually begins either with the sufferer trying to work out why they're so shattered all the time or with their partner complaining about the terrible snoring and snorting. Diagnosing apnea will involve tests during sleep, such as checking oxygen levels and monitoring breathing, heart and blood pressure during the sleep cycle.

If there is a diagnosis of sleep apnea, having five to fourteen epidodes of breathing interruption an hour only counts as a mild case. To count as severe, the poor distracted sleeper has to suffer at least thirty episodes an hour. Every two minutes there has be a temporary blockage in breathing.

But there are ways to treat apnea. Rather like snoring, the first line of defence is to cut out the factors that aggravate the narrowing of air passages – such as alcohol, smoking and lack of exercise. Losing weight can be the most direct route to reducing the problem. Breathing masks and air machines are also available to help people who otherwise face a dreadful night of gasping, breathless sleep. These are called 'continuous positive airway pressure' devices, and they operate by pumping air into a mask, which goes over the nose and mouth, with sufficient pressure to ensure that the airways remain open. Every night the sleeper puts on the mask and switches on the machine and by all accounts there can be a dramatic improvement. Sleep, so long denied, is available at the flick of a switch.

In some cases, sufferers are given surgery, particularly if the nasal passages are blocked or crooked in a way that makes it harder to breath.

Without a hint of irony, the NHS also has another official recommendation for apnea sufferers: take up the didgeridoo. Playing this Australian wind instrument with a comedy name is apparently a successful way of building up the body's breathing equipment. The NHS cites evidence that practising every night for four months can lead to a significant reduction in the problem.

But think about it. The poor partner, who has been driven crazy by this person's appalling snoring, then has to sleep next to someone in a plastic face mask. Now they want to play a didgeridoo every night for four months. When did a night's sleep become so complicated?

See also: The sleepless city, p. 142

Jet lag

ANYTHING that's enjoyable comes with a penalty clause. It's one of the unwritten rules of nature. So going on exciting long-distance adventures has to have a sting in the traveller's tale. International airports, with their endless queues and corridors, should be punishment enough. But there is also jet lag, a groggy feeling of disorientation that hits air passengers who have crossed too many time zones too quickly.

Jet lag is what happens when the body's circadian rhythms, its internal clock, get confused. The usual triggering of sleeping and waking is thrown out of step and the body doesn't know if it should be asleep or running around.

It is supposed to be easier to avoid jet lag when you travel from east to west, rather than west to east. The theory is that the body is better at adjusting to a longer day, which is what happens when you go west, when you have to put the time on your watch back. The body's internal clock stretches its day and is quicker to adapt to the new getting-up time.

But going east and putting your watch forward is a tougher challenge for the body's internal clock, which is still trying to fit in its own regular full-length day. Sleeping becomes extremely difficult and jet lag takes its toll.

So what can be done to lessen the impact?

Travellers can avoid exacerbating the problem by avoiding dehydration, alcohol, lack of sleep and stress.

Adjusting sleep patterns a few days in advance is also recommended, so someone heading east might begin to go to bed earlier, to lessen the impact of the changed time zone when they arrive. It is also worth considering when meetings will be scheduled, to avoid them being held in what your body still thinks is the middle of the night.

When you get to your destination, the advice is to get into natural daylight as soon as possible, as this will be a powerful signal for the internal body clock to readjust itself to the new surroundings. Eating times should also be those of your destination, to help you adjust to a new routine.

None of these good intentions will stop jet lag completely, so unsurprisingly there are plenty of theories and products that claim to have the answer. Among the most talked about remedies in the United States is the use of melatonin, the hormone that triggers a sensation of sleepiness. In Britain, it isn't licensed as an over-the-counter medicine, with the NHS saying the benefits are 'inconclusive'. There have been studies both supporting and challenging the effectiveness of melatonin remedies in tackling jet lag.

Plenty of herbal and homeopathic remedies are also on offer, along with ideas such as fooling the body by wearing sunglasses. But there has also been interest in adapting diets as a way of staving off the worst of jet lag. Former US president Ronald Reagan was among those who were thought to have used a pattern of alternate days of feasting and fasting before long flights. This anti-jet-lag diet was developed with and tested on US military personnel being transported across time zones, and it claims a high success rate in reducing the incidence of jet lag. On two feast days the soldiers could have an unlimited amount of high-protein food, while on the alternating two fast days they could only have 800 calories. After these four days, the disruption from the time-zone shift was claimed to be much reduced.

A recent study at Harvard University has also supported

the idea that fasting is the best answer. By denying food before and during travel, researchers found that the body readjusted more quickly to its new destination, with meals eaten on arrival at local time helping to reset the internal clock.

So don't eat, don't booze, avoid stress and avoid caffeine. Suddenly, staying at home seems much more relaxing.

See also: Weekend jet lag, p. 38

Restless leg syndrome

THIS IS a common sleep disorder that can be infuriating for both the sufferer and the person trying to sleep next to them. As the name suggests, it's characterised by an inability to stop moving the legs, particularly after a period of inactivity.

The sufferer has a sensation often described as itching or like something crawling over the skin, which they can only relieve by repeatedly moving or rubbing their legs. What makes it particularly annoying is that it's likely to strike when people are in bed and trying to rest. Sufferers have an irresistible urge to stretch, kick or shake their legs, to try to get rid of this irritation. When they stop moving their legs, the itching or burning sensation returns, stopping them from getting any proper sleep and leaving them fatigued the next morning.

It's usually associated with older people and with women during pregnancy, but it's possible for young people to develop restless leg syndrome too. Estimates of how many people suffer from this syndrome range from 3 per cent to

15 per cent of the population, with claims that the syndrome is often undiagnosed and under-reported.

There are a number of ideas about its origin. Iron deficiency is one possible cause, kidney disease is another, it's also thought to be a side-effect of some medications, including anti-depressants. Up to one in five women temporarily experience it in some form during pregnancy. There is also believed to be a considerable genetic influence, with many sufferers reporting that it had already affected someone in their family.

See also: Sleepwalking, p. 248

Snoring

ANY JOURNEY into sleep hell would have to involve listening to the reverberating sound of snoring. It is sleep torture for other people – the most common reason for couples sleeping apart, one of the worst culprits for ruining romance and one of the biggest disrupters of sleep. Anthony Burgess summed it up in these words: 'Laugh and the world laughs with you. Snore and you sleep alone.'

It is also one of the most common sleep problems, affecting as many as four out of ten men and three out of ten women. And animals suffer from it too. A twelve-year-old horse called Rocky in Aberdeenshire had to be operated on to stop a snore that was described as resembling a 'foghorn'. The horse's snore could be heard two fields away – and they have big fields in Aberdeenshire.

It can be desperately loud. The record holder, and it's a

questionable trophy to want to hold, is Kare Walkert of Kumla in Sweden, who clocked up ninety-three decibels. That's like trying to sleep next to a lawnmower.

It is also one of those ailments with a high embarrassment value. Waking up open-mouthed and snoring on the train might be bad enough, but how about the public humiliation of the person who got thrown out of a Steve Davis snooker semi-final for falling asleep and then filling the room with his snores? There is something grimly fascinating about seeing someone in such a ridiculous position, pulling such strange faces and generating such horrible noises. YouTube, that great freak circus of the digital age, is full of clips of people's nearest and dearest snoring away for all they're worth. Who wants to record these people, and who is watching this snore porn?

z z z z z z z z z z z z z z z z z

The record holder, and it's a questionable trophy to want to hold, is Kare Walkert of Kumla in Sweden, who clocked up ninety-three decibels. That's like trying to sleep next to a lawnmower.

Snoring is caused by the vibration of the soft palate and vocal cords and is often associated with some kind of narrowing of or blockage in the airways. It's noisy breathing, turning the nose, mouth and throat into a sound chamber. It is increasingly used as an identifier for the more serious complaint of sleep apnea.

A number of factors make snoring more likely. Drinking alcohol is pretty much top of the list, and most people will be able to attest to this particular by-product of a night on the beer. Snoring is aggravated by anything that disrupts breathing, which could be smoking, allergies or medication. Being overweight is another major culprit.

Tips to avoid snoring, apart from laying off the booze

and losing weight, include sleeping on your side rather than on your back. As a way of keeping people on their side, a suggestion is to attach a tennis ball to the back of the pyjamas, which discourages the sleeper from rolling onto their back during the night. It sounds odd but there are many claims that it works. You can even buy purpose-made 'anti-snore shirts' made with a lump in the back to train the sleeper into staying on their side.

There are also snoring pillows, designed to hold up the head in a position that keeps the airways open. Raising the head of the bed can also help to avoid the constricted breathing that leads to snoring, as can the use of nasal strips. Mouth snorers might like to try 'chin-up' tapes, which can be stuck on the sleeper's face in order to keep the mouth closed, so that you breathe through your nose. Probably not something to slip on for a first romantic night, though.

Phrases such as 'unobtrusive and easy to use' are a sure sign that something is going to look terrible, and there are all kinds of other slings and straps to be tied around the snorer's head and gum-shield contraptions to be wedged into their mouth. 'Nostril expander' is another phrase full of bedtime seduction. The ex-snorer might not disturb anyone any more, but chances are that, wearing all that equipment, they are unlikely to have anyone next to them in the bed anyway.

You can also buy a device to wear around the wrist that delivers a mild electric shock whenever it detects snoring – something that sounds rather like a training kit to deter a dog from straying. What would happen if two people were wearing them and they both snored? Can you imagine the arguments about who started the electric shock?

Physical exercise is recommended as a simple and effective anti-snoring tactic. It's likely to help with weight loss, and anything that opens up the airways is going to reduce snoring. Singing exercises have also been found to be a surprisingly successful way of strengthening the muscles in the throat and making constriction less likely. Research has been carried out in a hospital in Exeter to see if this could be used for more people with sleeping and breathing problems.

Snoring has never really had the respect it deserves, and is the subject of one of the oldest place-name gags in history. There is a village in Norfolk with the endlessly entertaining name of Great Snoring. In 1611, when the lord of the manor, Ralph Shelton, sold the village, he delivered the punchline: 'I can sleep without Snoring.'

See also Sleep apnea, p. 194

New parents: know them by their eyes

THERE'S an old joke about babies settling into a regular sleep pattern after the first few weeks – one hour asleep and twenty-three hours of furious crying.

Lack of sleep is one of the biggest shocks to the system when a first baby is born. It feels like you've given birth to an air-raid siren. Not only can babies demand attention when you're at your most sleep-exhausted, but they can keep it up for hours. Night after night, your gorgeous baby is going to stop you sleeping.

Saucer-eyed and groggy, you find that the days are

The lack of sleep can make new parents feel rather strange.

becoming a sleepless blur. It can feel like some terrible endurance test. The antenatal classes should have been giving lessons on life without sleep, not all that stuff about when to call the midwife. There might be loads of glossy, wipe-clean books with all kinds of advice about feeding times and developing sleeping patterns and not getting babies dependent on the parents' presence for sleep ... but in the middle of the night with a howling baby, none of that makes much difference.

Once the baby arrives the old pre-child life has gone for ever. Sleep will never be the same again. And the first few months of a baby's life are a tough initiation for the parents, the baby sleeping irregular hours while Mum and Dad still fondly believe they can return to their old routines. Babies might sleep for sixteen hours a day, but they might not be the same sixteen hours as anyone else in the house. The baby is breaking the parents in, teaching them that they are now obliged to follow the random and wrathful demands of their offspring.

What is really crushing for parents is that they still have to carry out a few tasks in the outside world, such as going to work. Any working parent will remember that awful moment when daylight is beginning to appear and they still haven't had any proper sleep. It's going to be one of those zombie days.

It's also difficult for a couple's relationship. It might mean sleeping apart temporarily, so that one person is alert enough for work, or taking it in turns to get up when the baby cries. Sleep deprivation isn't the best mood enhancer.

Efforts to soothe a baby at some unearthly hour can make for strangely surreal night-time moments. One of my daughters developed a strange fascination for baseball on television in the depths of the night. Maybe it was the bright green pitch or the grid of the markings, but night after night she watched this, howling if I turned it off. I don't even know the rules of baseball, but that summer I became its biggest fan.

Getting young children to sleep can become such a big issue that all kinds of rituals develop around it. The toy that helps children to fall asleep becomes the most important object in the house. Losing Mr Lamb or Olly Dolly or whatever this toy is called would be the most terrifying thing that the family could imagine. Secretly they might buy a spare. Then they might buy another spare in case the spare gets lost.

Certain bedtime books and nursery rhymes often seem to help more than others too, and these then become part of the midnight mind games, as you sing 'Horsey, horsey won't you stop' or 'The grand old duke of York' like some crazed person auditioning in front of the stern jury of your child. Get it wrong and there's a terrible penalty. No performer was ever more grateful to close the eyes of an audience.

See also Oversleeping, p. 75

Sleep training: quack alert

PICTURE THE SCENE. I'm lying on the floor, it's the middle of the night and all the lights are turned off. I've had about four hours' sleep in the last twenty-four hours but I'm still awake, singing a nursery rhyme. One arm is wedged through the bars of a wooden cot, so that one daughter can hold my hand; my other arm is stretched at another angle so that another daughter in her bed can hold on to my other hand. If I move so much as an inch in any direction they will both erupt into angry crying so loud and piercing that it will be audible in Canada. I'm not sure that this is entirely what was intended by the sleep training manual. Maybe we got lost somewhere on page 73 of the bossy, irritating, smug book filled with all that rubbish about how easy it is to get your child to become a happy independent sleeper in only a few days.

It always starts out sounding like such a good idea. It might be tough, and it might mean playing the hard-hearted parent for a few nights until the young child gets into a routine of being able to fall asleep unaccompanied and alone, but sleep training promises a way of setting rules and creating good habits that will help everyone in a family to get a good night's sleep. Why, then, is it always promoted by some bleached-teeth expert whom you secretly suspect of having bought all their qualifications on the internet?

The rules of sleep training, like elaborate diets that are never really going to work, demand absolute attention to detail. They take many forms, with stipulations about how long the child can be left crying on each night, and what the guilt-stricken parent is meant to say when leaving the room amid a cacophony of wails. But really, like stupid diets, they all take the same path to failure.

The parents, feeling dreadful and upset, always find themselves on the landing outside the door of the child's bedroom, looking at each other reproachfully, then peeking guiltily through the crack of the door into the bedroom where the child is meant to be sleeping. This is supposed to be about sleep training, teaching the child how to remain alone. Instead it breeds resentment and recrimination between the parents, who start secretly blaming each other. Whose idea was this in the first place? Whose control-freak friend recommended this sleep training by Dr Sleep Fascist with the Fake Tan and Fake Californian Qualifications? By now, the child or children are howling so loudly that the walls are shaking. They seem to have found some whole new level of anger and noise. International monitoring stations are picking up this volcano of unhappiness. Is it possible that a child could actually explode with anger?

Of course, it is the most natural thing in the world to go and comfort a child when they are crying. But Dr Sleep Fascist has ruled this out. This would be like giving an alcoholic a bottle of whiskey, or giving your credit card number to a compulsive gambler. The howling child must be left alone at all costs; they must be treated with the caution you'd apply if you'd stored rods of nuclear fuel in the bedroom. Do not enter, do not touch, step away from the bedroom door.

The screaming is now so loud that pictures have started falling off the wall. The poor child is entirely baffled as to why its parents, who he can hear swearing at each other outside the bedroom door, won't come in like normal. What kind of cruel mind games are they playing? Why are they skulking around in the dark upstairs rather than coming in to read stories, smelling of wine and complaining about work?

Eventually overcome with guilt for not going to the crying child and then filled with a sense of failure for giving up the sleep plan on the first night, the parents storm into the bedroom. It's like the end of some kind of hostage scene, the parents hugging the furious offspring, who seems to be on the verge of exploding into flames of molten fury. And all this has been in the name of sleep. It's going to be another bad night. It's going to be another exhausted morning. It's going to feel as though someone has been rubbing sandpaper in your eyes all night.

So I'm back holding hands in the dark to help my children go to sleep, stretched out on the floor like some kind of sleep-deprived martial artist, desperate for their eyes to finally close, knowing that the slightest sign of movement on my part will trigger crying louder than a car alarm in a concrete multi-storey.

zzzzzzzzzzzzzzzzz

Sleep training is about as natural as buying the groceries at three in the morning, it's a modern invention for a problem that never previously existed.

Sleep training is a way of making everyone feel bad about something that is perfectly natural. No one needs training in going to bed. Young creatures like to have their parents with them when they're falling asleep. It makes them feel secure. That's why people have been telling bedtime stories and singing lullabies for the past 5,000 years. Sleep training is about as natural as buying the groceries at three in the morning, it's a modern invention for a problem that never previously existed. Do animals need books about sleep training? Did human beings for tens of thousands of years require a quack in a white coat to tell them how to put their children to sleep?

Being a parent of young children is a tough time in anyone's sleeping life. You can see the evidence in the hollowed-out eyes of the new parent, stunned by the anti-sleep device to which they've given birth. How can something so small and sweet in the daytime turn into such an all-conquering monster at night, destroying sleep with an aggression and accuracy that almost seems supernatural? How do they know to wait until you have just fallen asleep to start crying again, snatching away any hope of rest at the exact point when sleep almost seems possible?

Denying sleep is the baby's way of taking control, making sure that the sleep-starved parents know that whatever was once important to them is now in the grip of their angry little fingers. There's no training to escape that.

See also Sleeping together, p. 94

Worst place to wake up?

I had retired. My wife called me. I was moaning in my sleep. My wife called me, and says, 'Wake up; you are dreaming;' and I was dreaming, and as I woke up I heard a slight crash. I paid no attention to it until I heard the engines stop. When the engines stopped I said, 'There is something serious; there is something wrong. We had better go up on deck.' I just put on what clothes I could grab, and my wife put on her kimono, and we went up to the top deck and walked around there. There were not many people around there. That was where the lifeboats were. We came down to the next deck, and the captain came up. I supposed he had come up from investigating the damage. He had a very serious and a very grave face. I then said to my wife, 'This is a very serious matter, I believe.'

THIS WAS THE TESTIMONY of Charles Stengel, a first-class passenger on the *Titanic,* the liner which sank on 15 April 1912 after colliding with an iceberg in the Atlantic. Mr Stengel and his wife, Annie May, from New Jersey, both survived the disaster.

See also Why do children like frightening bedtime stories?, p. 60

Moon mission: Much about sleep remains a mystery.

Mystery Ride

What is sleep for?

It's such a basic question, but there is no straightforward answer. Other fundamentals of life are much more self-explanatory. Why do we eat? To get energy. Why do we breathe? To get the oxygen we need. Why do we spend a third of our lives in sleep? Not quite sure.

Many suggestions can be filed under 'not entirely convincing'. Is it a way of improving survival, allowing us to lie low and out of the way of harm during the dangers of the night? If that was the case, then surely canny old Mother Nature would have come up with a better tactic than leaving us as vulnerable as a baby, unconscious for hours, unaware of any approaching predator.

Some have speculated that it could be about conserving energy, about shutting down for the night in an efficient way, like a cleverly timed central heating system. But this theory has also been dismissed, as we save only a small amount of energy this way. It's not as if the body and brain stop functioning during sleep, they just behave differently.

More recent research has looked at a different kind of purpose for sleep, to do with how we store memories, how we learn, how we create our sense of who we are, how our bodies repair themselves, how young people grow. It is when the body retunes itself to its natural rhythms. If that all sounds rather vague, then that's because researchers are still only in the foothills of understanding.

A much more definite picture emerges when we turn the question on its head. We don't know what sleep is for, but we do know what happens when we don't sleep. People become sick and emotionally exhausted, they hallucinate and the immune system crumbles. Rats completely deprived of sleep drop dead. What do they die of? Well, once more we have more questions than answers.

We also know that there is a price to pay when our sleep patterns are disturbed. We might not fully understand the meaning of the body's inner clock, its circadian rhythms, but ignore them too often and you'll end up feeling like you've been used in some dreadful chemical experiment. Night-shift workers will be nodding their weary heads in recognition.

So we clearly need regular sleep. Even though we don't know what it's for, we know we can't do without it. Meanwhile, the links between sleep and how we consolidate our experiences and memories are being busily explored. Researchers are also looking at the way that the different stages of sleep appear to serve different functions, rather like the way that different types of food and drink are needed for an overall sense of well-being. This research is still at a relatively early stage, but a particularly important area of study is REM sleep, which is being examined for its connection to memory. Human beings have been wandering the planet for a couple of hundred thousand years, but this 'rapid eye movement' sleep was only formally identified in 1953. Perhaps people had seen it before but had never given it a name, let alone named a rock band after it.

More to the point, almost as soon as REM sleep seemed

like an encouraging place to look for a meaning for sleep, more question marks seemed to arise. A man who suffered a brain injury and had no REM sleep at all didn't appear to have any memory problems. Anti-depressant drugs knock out REM sleep for large numbers of people, without any consequences for their memory.

z z z z z z z z z z z z z z z z z

Researchers are also looking at the way that the different stages of sleep appear to serve different functions, rather like the way that different types of food and drink are needed for an overall sense of well-being.

Researchers will continue to attempt to find what sleep does, attaching ever more elaborate devices to the heads of drowsy university students. Their findings are probably more reliable than the sleep experiments carried out on California's prison population in the 1930s. In one study, inmates were measured to see the effect of coffee on their sleep. The results were fairly inconclusive, and it's difficult not to think that this incarcerated collection of murderers, kidnappers and robbers had more worries to keep them awake at night than a cup of coffee.

There's certainly nothing new about puzzling over sleep. Scientists, philosophers and poets have been speculating about it for centuries. The fact that so many of their ideas now seem ridiculous is only a warning that our own theories are likely to sound just as absurd to future generations.

In Elizabethan England there was a flurry of scientific interest in sleep. Thomas Coghan, writing in his book *Haven of Health* in 1584, described the widespread belief that sleep involved a kind of evaporation of heat from within the dormant body, with the fumes rising upwards from the stomach. As plain proof he pointed to the way that people were particularly given to falling asleep after they had been eating and drinking. Elizabethans tended to regard sleep as a relation of death, an image of mortality in our everyday lives, a foretaste of eternal slumber. Shakespeare returned again and again to this theme. But more optimistically, writers of that age also saw sleep as a reflection of nature's kindness, a gift for rich and poor alike. This view was widely held. In Spain in 1605, Miguel de Cervantes wrote that sleep was 'meat for the hungry, drink for the thirsty, heat for the cold, cold for the hot'. It was a mysterious and manifold gift.

The sense of not really knowing what sleep is for goes back a very long way. In 350 BC, Aristotle in his essay 'On Sleep and Sleeplessness' makes some observations that are as profound now as they were then. Sleep engulfs the whole person, all their senses – it's not a partial shutdown, it's not the body or mind taking an independent rest. The sleeping state is as absolute as the waking state. Aristotle argues that both waking and sleeping are inseparable. Without waking there is no sleep, and without sleep there is no waking. It's not that we journey out of our lives into sleep, it's that sleep is as an inevitable and natural a part of our life as being awake.

Sleeping and waking are necessary halves of the experience of being alive. Asking the purpose of sleep might be like asking ourselves what is the purpose of being awake. Discuss. Another night.

See also: What happens when we fall asleep?, p. 103

The meaning of 'nightmare'

THE WORD 'nightmare' has bizarre sexual origins. It comes from 'maere', an Old English word meaning an 'incubus', or an evil demon or spirit that creeps in during the night and suffocates or presses upon the sleeper. 'Incubus' comes from the Latin meaning 'lie upon', and the demon was believed to lie upon human women and have sex with them while they slept. The word 'incubus' is now used to mean a source of stress or anxiety, but its older folk meaning spawned the word nightmare, the visitation of such demons during the night. This root of the word links the nightmare with feelings of powerlessness, suffocation and vulnerability during sleep, with sexual threat and the malignant supernatural.

See also Forty winks, p. 33

Dreamland

DREAMS CAN BE SURREAL, soothing or sensual. They can feel heavy with symbolism or bizarrely random. But dreams are in our own heads, they're uniquely our own creation – no one else will ever see them. In the feverish vision of *Heart of Darkness*, Joseph Conrad throws out the line 'We live as we dream – alone.' We are utterly alone with our dreams; they are our own inner world pitched against external reality. They're also spontaneous, created about four to six times every night and viewed without instruction

Bedroom furniture on sale in London in 1902.

or control. But once we've made this dreamland, we become its prisoner. We can't run away from the dreams we've made. And the next morning we're only likely to remember about one in twenty of them.

When we wake up from a particularly atmospheric dream, the mood can linger into the next day. It can feel like it must mean something. If something vivid happened in a dream, involving real people or events, it's hard not to think that this is some kind of message. But where do we get our dreams? What makes us generate these mind movies? It's like watching a very energetic, very confused piece of experimental drama. If there is a message, what is it?

As long as people have been dreaming, they've been trying to work out what dreams mean. Five thousand years ago, Sumerian scribes were recording dreams. The ancient Egyptians devised an elaborate set of dream interpretations. The ancient Greeks were divided between those who thought dreams could bring divine messages and those who thought they were just the jumbled reflections of something triggered during the day. These dream sceptics argued that dogs had dreams too and they were hardly being sent visions by the gods.

An intriguing insight into dreaming is that people from all over the world, in the most diverse of cultures, have the same types of dreams. A study in the 1960s catalogued tens of thousands of dreams and found that wherever or however people live, we all have dreams that follow the same basic scripts. Dreams might involve different cultural settings – you're not going to dream about turning up late for an exam if you've never been to school, for example. But universal themes include teeth falling out, appearing naked or exposed in public, forgetting the words to some kind of public speech, not being able to find something, meeting a relative or friend who has died, returning to a childhood home and missing an important appointment. Whatever it says about the human condition, the most common theme in our dreams is anxiety. Being chased, being unable to escape or get away quickly enough, arriving late and missing something important, making a fool of ourselves in public . . . these scenarios all seem immediately familiar. Only a tenth of dreams have any sexual content – and these are most likely to be experienced by teenagers. Another example of what's wasted on the young.

The psychology department of the University of California, Santa Cruz, has a 'Dreambank' of more than 16,000 records of dreams. If you dip randomly into this collection of experiences, from all kinds of different people, the most striking factor is how familiar they seem. There are dreams of wearing inappropriate clothes, dreams about threatening strangers, about remembering an old house, lots of dreams about friends and relatives and about familiar places that have been inexplicably altered. There are all kinds of specific local details, but a strange sense of a shared experience.

Another common factor is that we all tend to feature in our own dreams. They might be confused and incoherent, but we are usually the central character trying to make sense of them. And what often makes dreams seem so strange is the way that time is distorted and scenes fail to follow a predictable sequence. How long does a dream last? Dreams, roaming across our lives and memories, step out of time.

Individuals also have dreams that come round and round again. These are not quite recurrent dreams, where exactly the same thing happens each time, but dreams in which the same things are constantly recycled. A study of a woman in the United States who kept a diary of her dreams for over fifty years showed that three-quarters of her dreams were from a small range, for example dreams of losing her purse, of people breaking into her room, dreams about her mother and of being late or missing a bus.

That isn't to say that dreams aren't driven by specific events. Upsetting moments are often replayed in dreams. In fact, anxiety dreams are a defining feature of post-traumatic stress. Extremely vivid, often terrifying dreams characterise this condition, with the sufferer re-enacting the stress event in dream after dream. Vietnam veterans can relive the same moment of terror month after month, year after year. But in many ways such harrowing dreams, triggered by real experiences, are the exception. Scientists have had little success when they've tried to manipulate what we see in dreams, for example depriving someone of water in order to see if this generates dreams about thirst or drinking, or showing vounteers violent or erotic films before sleep. In one experiment, participants had their eyes taped open during sleep and were shown images and objects, again without any of these appearing in their dreams. It seems that dreams belong in another world. There is no simplistic link between the outside world and the inner dream.

The dream might be an elusive creature to catch, but since the discovery of REM sleep in 1953 it has been possible to say where it is likely to take place. It was soon discovered that when people were woken up in this stage of sleep, they were much more likely to be able to recall their dreams. It wasn't the only stage of sleep where dreams were reported, but it was by far the most common. REM sleep has also been linked to the processing of memories, and there have been suggestions that dreams might be related to this in some way.

One rather appealing theory is that dreams are the play-back of our long-term memories, always present, which are always whirring away in the background, but are suppressed during the waking day and only visible in the stillness of sleep. It's an extension of the idea that dreams are a snatched glimpse of our secret inner lives, revealing our concealed emotions.

As the psychiatrist and dream interpreter Carl Jung said, 'Who looks outside, dreams, who looks inside, awakens.'

See also Does cheese give you nightmares?, p. 135

REM sleep

IN THE 1950s, researchers at the University of Chicago made one of the great breakthroughs in understanding sleep when they identified a separate stage of sleep characterised by 'rapid, jerky eye movements'. Whatever was causing this movement in the sleepers' eyes also appeared to be connected to increased brain activity. The brain seemed to be as busy as when awake, and another piece in the jigsaw was the discovery that anyone woken from this stage of 'rapid eye movement sleep' was much more likely to report that they had been dreaming. There seemed to be a strong connection between entering this stage of sleep and having dreams.

The discovery of REM sleep changed the way that scientists approached the study of sleep. It showed conclusively that sleep was not simply a passive state of not being awake, but instead a complex, changing experience in its own right. It also emphasised the need to study sleep as a physical event. For centuries, sleep had been analysed by philosophers, writers and psychologists, but now scientists had evidence that sleep and dreams needed to be studied as a physical phenomenon. During REM sleep there are changes in brain activity, the muscles are temporarily paralysed and the body's temperature and heart rate become irregular.

But sleep still remains a mysterious creature. The more that's found out about it, the more questions are raised. Once the existence of REM sleep was established it was then discovered that the young have much more REM sleep than the old. Does this mean that the old really do run out of dreams?

Newborn babies can sleep for sixteen hours a day, and about half of that time is REM sleep. When you look into the cot at a young baby, you are looking at someone who experiences the equivalent of a full working day of REM sleep. In adults REM sleep might only constitute a quarter or a fifth of sleep; the sleep deprived can have even less. This means that a baby could be spending four or five times longer each day in this dream state than its parents.

zzzzzzzzzzzzzzzzz

For centuries, sleep had been analysed by philosophers, writers and psychologists, but now scientists had evidence that sleep and dreams needed to be studied as a physical phenomenon.

There is something almost alien about a newborn baby, as though they have not yet entirely materialised, as though they are as much of the world they have left as the world they have entered. Maybe all this time in REM sleep, spaced out in dreams, might have something to do with the way they can seem to be so curiously distant, as though thinking about returning to some other planet. They are weak and dependent on their new parents, but their thoughts are roaming across their own inner dreamworld. It is believed that babies in the womb spend an even higher proportion of their time in REM sleep. What experiences does a child in the womb or a newborn baby use to make dreams? What have they seen to dream about for so long?

There are other uncertain connections between the inner world of the mind and REM sleep. People suffering from depression enter REM sleep much more rapidly than the average sleeper, and many anti-depressants are deliberately designed to delay and suppress REM sleep. This interference with REM sleep is a direct way of manipulating the mood and emotions of the patient.

What should we make of all this? The newborn infant seems to need huge amounts of this type of sleep as it enters the unfamiliar outside world, while the depressed adult takes drugs to prevent it happening at all and this seems to improve their sense of well-being. The way we sleep is quite clearly interwoven with our mood, personality and how we experience the world around us.

See also What happens when we fall asleep?, p. 103

Freud and Jung

LOOKING FOR the meaning of dreams is an ancient obsession. From the earliest times people have seen dreams as coded messages, as something supernatural brought from outside the human world. And opposing such ideas, for almost as long, have been the rationalists who have seen dreams in a more prosaic light, who view dreaming simply as something functional that happens during sleep – a bit like snoring only more interesting. The rationalists believe that dreams arise from the everyday world – that they are distorted projections of experiences from the waking hours, not a vision from another dimension.

Then Sigmund Freud came along. His psychological interpretation of dreams, written at the very end of the nineteenth century, created an entirely new way of looking at their origin. He saw them as reflections of the inner struggles and dramas of the inner person, carrying the truths that the waking self did not want to acknowledge. Dreams were not random or bizarre, they were symbolic

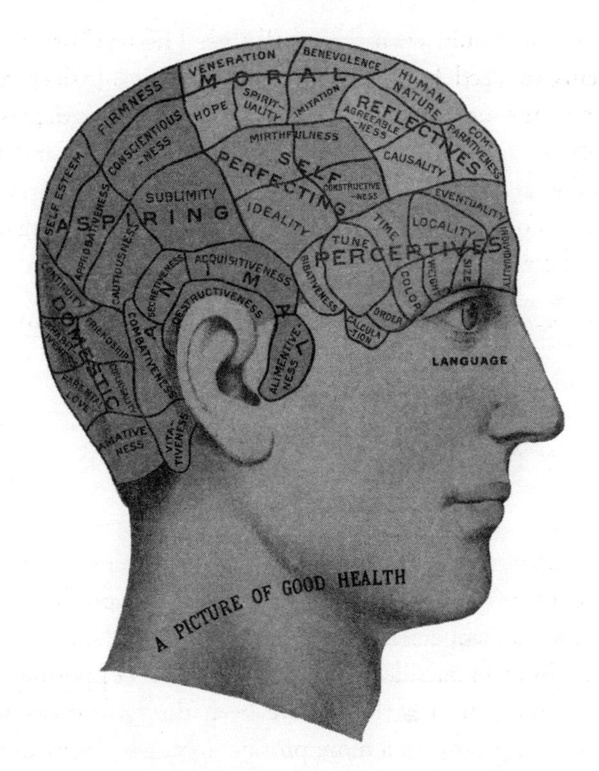

A map of the 'animal' and 'moral' brain from 1870.

and very specific workings out of experiences and anxieties. The dream wasn't a vision sent by an outside force, it was an expression from within.

In particular, Freud believed that the sexual anxieties and frustrations that are not talked about during the day appear in the imagery of dreams. Dream symbolism is therefore the sleeper's way of exploring their sexual natures. The repressed emotions and the feelings that the waking person might not want to recognise in themselves are set loose in dreams, in images filled with a hidden meaning. Below the surface of the story of the dream are deeper, 'latent' messages. This is the work of the unconscious, the darkest corners of the mind having their own private conversation. The dream offers a place where experimental wish fulfilment can be carried out, wish fulfilment that is impossible in a respectable waking life.

Freud argued that the meanings of dreams are conveyed in different ways. Rather than confront worries directly, the dreamer might use alternative 'displaced' images that make the subject feel safer. Freud became famous for regarding dreams of pointed upright objects and various types of caves, entrances and assorted fruits as the dreamer's way of dealing with sexual desire. The dreamer might also project their own hidden feelings on to other characters in their dreams, these characters acting out the behaviour, perhaps not very nice behaviour, that the dreamer is inwardly considering – that violent maniac in the dream might be taking the rap for the dreamer's suppressed aggression, for example.

Freud also argued that a single action or object in a dream might have more than one meaning. Dreams are drawn from a long bank of experience and are not simplistic or one-dimensional.

In the century since he published his first works on dreaming there have been challenges to Freud's interpretations. But there is no denying the way he changed the landscape of how dreams are seen. Salvador Dalí's dreamscapes, with all those melted clocks, and Luis Buñuel's surrealist movies inhabit a territory that Freud had pioneered.

Following in Freud's footsteps, although they later fell out, was Carl Jung, the Swiss psychiatrist who took forward the concept that the images and stories which appear in dreams represent bigger underlying ideas. 'The dream is a little hidden door in the innermost and most secret recesses of the soul,' he said.

While Freud looked inwards, into the world inside each individual, Jung also wanted to look outwards, at the way that dreams and dreaming connect us to each other. He developed the idea of the 'collective unconscious', a shared understanding between humans of all kinds, something we all inherit, which becomes apparent in dreams. He linked this to the mythologies and folklore stories of different cultures, which also have common patterns and symbols.

Inhabiting this shared dream landscape, Jung believed, were archetypal characters, elemental figures that appeared in different guises, but were common to dreamers in different cultures and eras. These include the mother, the father, the child and the hero. But these figures might be expressed in different ways. For instance, if someone dreams of the archetypal figure of the mother, such a dream might not represent the dreamer's own mother. Instead, other associations and emotions might be explored through this archetype, such as thoughts about going back to a childhood home or fears about separation.

The individual might have different faces in this dreamscape. There might be a public 'persona', a figure that represents the dreamer in their own dreams. There might also be a darker, unrestrained 'shadow', reflecting the parts of a character that are usually repressed or feared. Jung also believed that men and women have both male and female aspects, and that we can be represented in dreams by either our male or female qualities.

Jung was a free-ranging thinker about dreams, dipping into science, culture, religion, mythology and art. But he believed that a dream was not a kind of play-acting of frustrated desires, but a faithful representation of the person having a dream. It was a little snapshot of their psyche. 'The dream shows the inner truth and reality of the patient as it really is: not as I conjecture it to be, and not as he would like it to be, but *as it is*.'

See also Surrealism and dreams, p. 235

Recurring dreams

SOMEONE IS COMING, the unseen shape of a threatening figure, you can hear their footsteps, you can hear them getting closer, but when you try to run away, you stumble and then struggle to get up again. They're getting closer now, you can hear their breathing, but your legs don't seem to be working properly . . . you've lost all your strength just when you need it most. They're going to catch you now unless you run . . .

Recurring dreams, which are more likely to be frightening than uplifting, are very common. They're not to be confused with trauma dreams, where people in their sleep re-experience a terrible moment that has happened to them in real life, such as the night terrors faced by survivors of the First World War. The recurrent dream might concern something that has never happened to the sleeper, but which keeps returning in dreams in more or less the same form. It might be something vague or unexplained, such as running through the woods trying to escape an unseen figure, or it might involve a more complex sequence of events, with more of a script and characters, which is endlessly replayed.

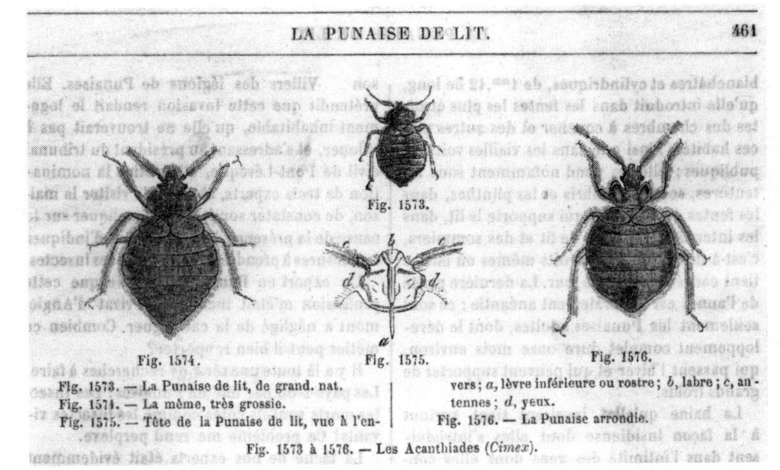
LA PUNAISE DE LIT. 461

Fig. 1573.

Fig. 1574.

Fig. 1575.

Fig. 1576.

Fig. 1573. — La Punaise de lit, de grand. nat.
Fig. 1574. — La même, très grossie.
Fig. 1575. — Tête de la Punaise de lit, vue à l'envers; *a*, lèvre inférieure ou rostre; *b*, labre; *c*, antennes; *d*, yeux.
Fig. 1576. — La Punaise arrondie.

Fig. 1573 à 1576. — Les Acanthiades (*Cimex*).

Sleep tight and don't let the . . .

You can often feel a sense of panic when you wake from one of those Nazis-have-found-me-hiding-in-the-cellar dreams, and it's not difficult to make a symbolic connection between these dreams and other anxieties in our lives. It's just that these dreams always seem to make the point in the most melodramatic and often violent way possible.

Recurring dreams are not necessarily with us throughout our lives. They might only occur for certain periods of time. Adolescents can have very vivid and frightening recurrent dreams that either disappear when they get older or surface only during anxious times as adults. A study in Australia found that students in the build-up to exams were more likely to have recurrent dreams. Once the stress recedes, so does the likelihood of the recurrent dream. The big bad wolf that has been frightening us since childhood slinks back to the woods.

Why do we haunt ourselves with such deranged scripts?

Are these primeval terrors lurking around the caves of our modern minds? What are we looking for in these bleak stories?

Psychiatrists have seen this type of repetitive, self-scaring dream in two distinct ways: as a subconscious attempt to address an unresolved emotional conflict, some unfinished business from the past that still upsets us; or as a comforter, a way of confronting an old fear and so taking control of it. It's a bit like the way children like to go back to the fairy tales that have most scared them. When faced by a new and unfamiliar threat in the outside world, we have old nightmares in order to see how we survived them.

The odd thing about recurrent dreams is that even though we feel that we are unble to control them, no one else has made them: they are entirely our own creations. Do we get the recurrent dreams that we deserve? Behind the trappings of adulthood, is that really how we see ourselves – as a child locked in a room, or a man always about to fall off a ladder?

Here are five typical recurring dreams:

Being chased

Looking over your shoulder, hearing footsteps in pursuit, legs turning to jelly, being hunted by a dangerous animal, a hand creeping round the door – yes, we've all been there. It's not always clear who or what is doing the chasing, but it's something that fills us with fear. It's almost as if all our emotional and physical terrors have been distilled into this one unnamed figure. Everything that has ever worried us is closing in, and we're trying desperately to keep ahead.

Falling

A cliff, a bridge, a ladder, a ledge on a high building – these are all classic dream settings for a sudden fall. You lose your balance, hurtle downwards, terrified, and just as you're about to hit the ground you wake up.

Flying

This is a much more upbeat dream, the sensation of flying making you feel powerful and in control. No one ever asks mid-dream how it is that flying through the air is so easy, but it's too much fun to stop. You are literally spreading your wings. Of course, dreams of flying in an aeroplane that is about to crash, with everyone around you screaming in panic, are not quite so uplifting.

Exams

Turning over the exam sheet and finding all the questions are unanswerable, failing to find the exam hall, beginning a speech at school and then forgetting the words, a frightening teacher – these are all angst-inducing throwbacks to youthful stress. These unsettling dreams are about facing up to scrutiny and being judged, revealing our deep fears of being unprepared or inadequate.

Trapped

You might feel suffocated, or dream of drowning or of a locked room that you can't escape, or experience that awful sensation of wanting to run but feeling paralysed. Or you might dream of being separated from loved ones, of being left behind or unable to protect yourself from an approaching danger. All these dreams are accompanied by horrible feelings of vulnerability and isolation.

See also Dying for sleep, p. 167

Heroes under the hill

IF SO MANY different individuals in different cultures have the same kinds of dream, it shouldn't be surprising that so many countries share the same kinds of folk stories. These stories have recurrent themes, like the recurrent dreams of sleepers, which take different forms but are essentially telling the same tale.

Among these classic folk stories is the tale of the hero who is sleeping and who will return at a moment of crisis when he is most needed. There are hundreds of versions of folk tales involving sleeping heroes, kings and princes. Typically, a peasant or a curious child will stumble upon an opening into a cave or below a hill and there find, in shining armour, a slumbering king. The king might wake for a second and ask if he is needed now, and when it seems that there is no urgent danger the king will return to his long, dusty sleep. The next time anyone looks for the entrance, it will have disappeared.

This is an evocative and romantic notion. The hero is not dead and not forgotten, but waiting in enchanted sleep, ready to return and inspire his people when their need is at its greatest. The most familiar version concerns King

Arthur, the archetypal sleeping hero under the hill, whose myth is now a strange fusion of Celtic roots, medieval chivalry, new-age tea shops and Victorian pre-Raphaelite imagery. After his final battle, the warrior king went to Avalon, where, according to tradition, he did not die. Instead he rests in a perpetual sleep, waiting for the time when he will be called upon to rise again.

The sleeping hero is there for us like the parent we thought we'd lost, ready to intervene at the very point when all hope is gone. Arthur and his knights are out there sleeping somewhere in the heart of a hill, waiting patiently while the centuries pass.

It's a peculiarly haunting myth, which has been reworked endlessly. Medieval writers gave it an intense spiritual glamour, the Victorians the melancholy of lost innocence and fallen nobility, and in the twentieth century it was linked to ideas of national survival. In the United States, Camelot was adopted as a metaphor for the gilded, doomed presidency of John F. Kennedy.

Myths have a powerful instinct for their own survival, jumping aboard any passing vehicle, remaking themselves in the image of the age. But there is something stirring about the phrase that is supposed to be written above the entrance to the hill where Arthur is sleeping so soundly. Out there, among the motorways, the sprawling towns and the supersized shopping centres, there is a hidden entrance to the heroic. 'Here lies Arthur, the once and future king.'

See also Fairy-tale ending, p. 48

There is a legend that King Arthur is sleeping and will return again.

Surrealism and dreams

When Sigmund Freud unlocked the door to dreams, he ushered in some very odd art. Freud saw dreams as the place where repressed primitive instincts about sex and violence were free to roam. This was perfect material for painters, poets and film-makers who wanted to challenge the tight-corseted bourgeois world and create a new sense of modernity.

The surrealists borrowed the vocabulary of dreams, the dislocations, the confusion of time and strange symbolic events. André Breton, who wrote the surrealists' manifesto in 1924 and who had trained as a psychiatrist said that these radical artists were asserting the 'omnipotence of the dream'.

But how do you create art that makes you feel as if you're in a dream? How about making a cup and saucer lined with fur? That was one approach from the artist Meret Oppenheim. Or perhaps a pair of high-heeled shoes trussed up like a roast chicken. Victor Brauner produced a table that had the head and the tail of a stuffed fox, while André Breton experimented with ideas such as automatic writing, where words were produced with as little conscious control as events in a dream.

The painter Salvador Dalí took Freudian ideas about dreams and used them as a way of creating his own inner landscapes. His *Persistence of Memory*, painted in 1931, showing clocks melting against a strange landscape, is the language of dreams turned into art. This wasn't just a painting style, it was a lifestyle. When Dalí attended the International Surrealist Exhibition in London in 1936, he delivered a lecture wearing a diving suit. A lack of air threatened to bring a more final ending to his presentation than he had planned, and the helmet had to be wrenched off by a surrealist poet wielding a spanner. Dalí was fortunate. There's not always a surrealist poet with a spanner immediately available.

Dalí collaborated with the Spanish filmmaker Luis Buñuel, who used the dream sequence in films as a way of depicting the secret sexual and violent desires lurking below the surface of respectable middle-class life. His films, from *Un Chien Andalou* in 1928 through to *Belle de Jour* in 1967 are an eruption of frustrated sexuality and bizarre rituals, shot through with the sense that nothing really adds up. The dream was the perfect way of giving a shape to something that was about a disruption of order and conventional morality.

What were the artistic legacies of the surrealists and their love of dreams? Was it the black humour of Samuel Beckett's empty landscapes? Or the psychedelic pop artists of the 1960s?

How about this plot instead? A pig flies a plane to Africa to look for a rare bird in a story that involved characters with three heads, set against backgrounds that have an eerie resemblance to Dali's paintings. It's *Porky in Wackyland,* created in 1938 by those radical surrealists, Looney Tunes. A colour version of the cartoon was produced after the Second World War under the supervision of another artist, Fritz Freleng, who became one of the creators of that most surreal of heroes, Bugs Bunny.

If you want to complete the circle, then how about looking to Bugs Bunny's appearance in *The Big Snooze*. In this 1946 classic, Elmer Fudd walks out of the cartoon, but when he falls asleep, Bugs Bunny enters his dream to bring him back to his day job as the fall guy. Even by cartoon standards, it's weird, with multiple layers of reality, references to dream interpretation and gags about sleeping pills (which were initially censored).

As they always meant to sing in the Coca Cola adverts, it's a surreal thing.

See also The Big Sleep: the best sleep movies, p. 28

Dream poetry

THERE'S NOTHING more tantalising than the dream that you can't quite remember. It was there as you were waking up, a shimmering moment of perception, a phrase on the tip of your tongue. That dazzling first line for the novel was there for a second or two and then disappeared.

Showing there can be poetry on screen, the movie *Hook* (1991) captures the idea of this transitional state beautifully when Tinkerbell tells Peter Pan: 'You know that place between sleep and awake, the place where you can still remember dreaming? That's where I'll always love you, Peter Pan. That's where I'll be waiting.' Or, as Samuel Taylor Coleridge put it, the lost words of a poem dreamt and forgotten had 'passed away like the images on the surface of a stream into which a stone has been cast'.

Coleridge was the author of the dream poem 'Kubla Khan', which he said had been delivered to him in his sleep in 1797. The poet had been reading a book with the lines 'Here the Khan Kubla commanded a palace to be built' when with a little medicinal help, most likely opium,

Coleridge had fallen into a deep three-hour nap. During this time he says the poem 'rose up' before him, a composition of 200 or 300 lines which appeared in his dream without any 'sensation or consciousness of effort'.

When he woke up he began to scribble down this poem, trying to capture as much of it as possible, still in a state of dream-fuelled concentration. But then came a knock on the door. It was, wrote Coleridge, 'a person on business from Porlock'. When the poet returned to his work, the memory of the dream writing had evaporated.

The surviving poem, an exotic and incomplete fragment, a quarter of the length he claimed for the original vision, was described by Coleridge as a 'psychological curiosity'. Some have questioned whether this story was a kind of artifice in its own right, a frame devised for a beautifully unfinished work about nature and art. But set that aside, imagine the book falling from your own hand, eyes drowsy, the lush sound of the words ... this is how it begins:

In Xanadu did Kubla Khan
A stately pleasure-dome decree:
Where Alph, the sacred river, ran
Through caverns measureless to man
Down to a sunless sea.

So twice five miles of fertile ground
With walls and towers were girdled round:
And there were gardens bright with sinuous rills,
Where blossomed many an incense-bearing tree;
And here were forests ancient as the hills,
Enfolding sunny spots of greenery.

See also Sex, drugs and overdoses, p. 173

Dreams from the 1930s

IN 1937 an innovative experiment in social history was created by a group of anthropologists, writers, film-makers and artists. They wanted to record their times through the voices and everyday experiences of hundreds of individuals. This project was called Mass-Observation, and their archives comprise an unexpectedly moving account of the stuff of ordinary life.

In the first wave of this information gathering, people were asked to record their dreams from the night of 11 May and the morning of 12 May 1937, which was the day King George VI was crowned. These were not laboratory or special test cases, they were ordinary men and women who simply wrote down what they dreamed.

These dreams from more than seventy years ago show that while the contemporary details may change, the psychological themes remain the same. There are still anxieties about embarrassment, strange confrontations and missed connections. One of the participants in Norfolk recorded:

> I was back at school (a mixture of school and Cambridge). My class was to be taken by the Bishop of Norwich. I could not remember in which classroom, or where the classroom was and this worried me very much. I had not got my undergrad's gown or my bicycle, I had left them in college. I had to walk miles to get them as my wife had taken the car. I gate-crashed into a cocktail party. I remember saying to the hostess, 'I am afraid I have come here without an invitation, but I thought the chief constable was here.'

Another dream reflects the fact that the Spanish Civil War was being fought at the time.

> I met General Franco by the side of a lake. He said to me 'I have taken Bilbao.' I said 'Oh dear! I have just repainted the whole place.' He replied menacingly: 'Do you know what I am going to do? I am going to scrape off all your new paint.'

And this dream, from Lancashire, captures the perennial sense of things going wrong in dreams and then becoming very strange indeed.

> I sat in a meadow and what I imagined to be a cow came up to me and I seemed unafraid. It turned out to be a bull and tossed me in the air. I somersaulted and fell on my feet and was caught again by the bull. This continued many times until I was finally tossed high over a fence and awoke before I fell. I remember falling asleep again and very vaguely remember talking to a member of our clerical staff who has been deceased for eighteen months, he was talking derogatorily of the work of a certain person, who happened to be himself.

There was also some plain weirdness from Buckinghamshire: 'The man called me to see a strange sight in a flower-bed, a tortoise attacked by a pigeon.' Yes, we've all been there.

Best of all, there is some excellent period-detail pretension from a student at Cambridge, who manages to encapsulate political fashion and some voguish music in a single dream: 'I dreamt of having breakfast, while I sang the International in "Swing Time" (how, I don't know).'

See also Smokers' guilty dreams, p. 161

A soothing glimpse of the west, on sale in Britain in the 1930s.

Dream believers

IT MUST MEAN SOMETHING. You keep having this really vivid dream. What is it trying to say? It seems too deliberate and strange to be accidental. All the way back to that Mesopotamian best-seller, *The Epic of Gilgamesh,* written more than 3,000 years ago, there are records of people trying to interpret dreams. The Bible is full of people falling asleep and then receiving instructions in a dream vision.

The ancient Egyptians, always fascinated by the links between the worlds of the living and the dead, the mortal and the divine, were particularly interested in dreams. They began to record their interpretations in 'dream books', listing whether a dream was a good or a bad sign. A surviving Egyptian dream book from about 3,200 years ago is held in the British Museum. It includes the observations that 'if a man sees himself in a dream looking out of a window, good; it means the hearing of his cry.' But 'if a man sees himself in a dream with his bed catching fire, bad; it means driving away his wife.'

Other ancient Egyptian dream books record stranger messages. A woman dreaming about giving birth to a crocodile was apparently good news, as it was a sign of a big family. Perhaps less peculiarly, if you dreamt of a big cat, that was good news because it was seen as a sign of a really good harvest. A deep well wasn't such a reassuring omen as this was seen as a premonition of a stretch in prison. A dream about a mirror by an Egyptian man was a sign that a second wife was imminent, which presumably was one of those dreams that made for an interesting chat over breakfast.

Dream interpretations could take a more complicated turn, for example by playing on similar sounding words. Dreaming about eating a donkey was lucky, because the word for donkey sounded like the word for great.

This process of archiving and codifying dreams was taken a step further by a Greek writer, Artemidorus, living in the second century AD, who compiled the first dictionary of dreams. Ancient dream interpreters like Artemidorus were interested in dreams as a way of divining the future, so rather than believing that dreams revealed something about the individual, they viewed them as providing clues to future events. It was about projecting forwards, not reflecting backwards. For instance, a perfume seller who dreams about losing his nose is going to lose his business; a man who dreams his staff is broken will lose his health.

Artemidorus also examined the idea that the same dream could predict different events, depending on the character of the dreamer. A well-educated woman who dreams of giving birth to a snake could end up with a child who is a slippery but successful public speaker, whereas a bad woman who has the same dream could give birth to a child who becomes a robber.

The logic can get a bit baffling. Dreaming of a cat is a warning of a threat from an adulterer, because a cat is a bird thief and women are like birds. Obviously. And if a slave dreams of having three penises, it's a good sign for all concerned, because the slave is going to be freed and acquire two extra names in honour of the man who set the slave free. What would Freud have made of the same material?

Interestingly, there is a chance to compare, because Artemidorus looked at one of the all-time classic dreams: teeth falling out. He listed a range of interpretations,

A reclining Buddha statue in Thailand.

including different meanings attached to the loss of particular teeth. For instance, a dream about losing your incisors could represent either the loss of a young relative or the loss of treasure.

These interpretations all assume that the dream is an external warning, the loss of teeth being a code for something that is going to happen. This is very different from modern dream interpreters, who would see the dream as a projection of inner fears. So the falling out of teeth could reflect repressed worries about growing old, a sense of powerlessness or concerns about a loss of appearance. It could be about the loss of sexual identity and the approach of middle age. Freud linked it to fears over castration and impotence.

The idea that a dream is carrying a message raises other, bigger questions. Where is such a message coming from and why should we assume the intentions are benign? For instance, one First World War veteran recalled in his memoirs how he had been sleeping in the trenches when he had a powerful sense of danger from a dream. When he woke, he took the dream as a message and moved away, just in time to avoid an incoming shell which would have killed him if he'd stayed in place. This soldier, who considered himself saved by the intervention of a dream, was Adolf Hitler, then a corporal in the German army.

If dreams *are* telling us something, who is in charge of the script? Who decided to save Hitler and not someone who was more deserving? If dreams are completely random, it still raises the question of how the content of these random images is determined. If dreams are entirely meaningless, then why do we have them at all? If they are generated from within, authored by our own intentions, then why do we feel so out of control in our own dreams?

The more you tug away at the phenomenon of dreaming, the more questions you raise. Dreams might be spontaneously delivered, in the sense of occurring without any deliberation direction, but they are not produced from outside of ourselves. We are their authors and we consume them. If we spoke to ourselves, dreams would be our words. So why aren't they more enjoyable?

If ever there was a place where we feel that life is lived on different levels, it must be the dream. This is a meeting place with another, hidden part of our identity, something below the surface of everyday conscious reality. Each time we wake up from a dream, we're left with that feeling of an encounter with the unexplained.

See also What happens when we fall asleep?, p. 103

Circadian rhythms

SLEEP IS AN integral part our lives. It's something that connects us with the patterns and pulses of all kinds of living things. The internal clock that governs this cycle of sleeping and waking is known as the 'circadian rhythm,' with 'circadian' meaning 'approximately a day'.

This approximate day of about twenty-four hours is the frame around which the pattern of sleeping and waking is shaped. Plants open and close their leaves in time with this biological clock, fruit flies become active or lazy depending on the pulse of this great timekeeper, human beings wake up and fall asleep in keeping with the settings of this inner rhythm.

Circadian rhythms are a profoundly ingrained part of life in animals, whether it's an office worker or a mammal in the arctic wastes. This rhythm ticks away, our bodies changing as the daily cycle turns, getting sleepy, getting roused, getting hungry. The body's temperature, the hormones that are released, drowsiness and alertness are all connected to it, and it in turn is linked to the rise and fall of the sun and the changing of the seasons. Without getting too tree-hugging about it all, it's a link between humans and the fundamental patterns of nature. As the planet turns and day becomes night, so our body clock ticks away, preparing for the approach of sleep. Signals from the rising and falling levels of daylight trigger responses in these inner clocks. We might put on pyjamas and pull up the bedcovers, but really it's not that different from plants furling their leaves for the night.

Like the line in the blues song, 'you never miss your water till your well runs dry', we only really notice our circadian rhythms when we try to mess around with them. Jet lag or the exhaustion from shift work are the result of trying to override the natural settings of the biological clock. These rhythms are deep rooted; they're not something that we can just reset, like moving the hands on a clock. Experiments have shown that animals, including humans, when kept in isolation and away from any external cues about the passing of time, such as changes in light, will still follow the basic 24-hour cycle. It seems to be part of our physical make-up.

How much it is ingrained into us is the subject of much study, with some suggesting that this circadian clock is part of every single cell in our body. How would each of these multitudes of clocks connect and synchronise with each other?

And when did this clock start ticking inside of us? According to recent research, using tests on shoals of obliging zebra fish, this internal clock might be set in motion before birth. The starting point might be the mother's genes, like an ancient clock being passed on through the generations.

There are also tantalising questions about how this inner clock relates to the seasons. It's part of the shut-down process for animals that hibernate, but what seasonal patterns are ticking away in the body clocks of humans? Are we programmed to sleep differently in winter and summer, even though our working lives make no concessions for such changes?

In one intriguing experiment, human volunteers were kept in conditions of simulated summer and winter, without access to a watch to tell them when to sleep or wake, and it was found that in the summer, people took their sleep in a single stretch. However, in winter conditions of more sustained darkness, another pattern emerged. When left to devise their own pattern, people slept longer, but in two stretches, broken by a period of waking during the night. It sounds remarkably similar to the tradition of a 'first sleep' and then a second 'morning sleep' that would have been familiar across the villages of pre-industrial Europe.

Rather like the underground rivers that lie beneath modern cities, circadian rhythms are still there, out of sight and often out of time with our unnatural and overcrowded lives.

See also The bat's four-hour waking day, p. 138

Sleepwalking

SLEEPWALKING is a phenomenon that occurs when someone who is in the deepest stage of sleep gets up and walks around or carries out a simple task without waking up. It isn't dreaming or the acting out of dreams, but the deeply sleeping body operating on its own autopilot.

When someone is in the dream stage of sleep, their brain is buzzing with activity. But the sleepwalker takes their nightly stroll at the point when the brain is at its most drowsy and unplugged. The sleepwalker might appear to look around them and be aware of their movements, but when they wake they are unlikely to have any memory of where they have been.

zzzzzzzzzzzzzzzzz

But in reality the biggest problem is that someone sleepwalking is extremely hard to wake up – they are in the lower depths of sleep, far away from the outside world, and bringing them round can take time.

Although adults can continue to sleepwalk, it's a behaviour most commonly associated with childhood and adolescence, with many growing out of it without any real explanation as to why they started sleepwalking or why they stopped. As well as sleepwalking, they might also be sleep talking or getting themselves a drink of water, all in this rather eerily disengaged state.

The sleepwalker can sometimes get their actions confused, for example by trying to go back to bed in the wrong room or, more embarrassingly, going to the toilet in the wrong place, such as the corner of a bedroom. There's a persistent myth that waking someone up in this condition is dangerous, as if they were going to drop down in a heap like a vampire struck by daylight. But in reality the biggest problem is that someone sleepwalking is extremely hard to wake up – they are in the lower depths of sleep, far away from the outside world, and bringing them round can take time. Of course, if a sleepwalker is at risk, for examply by picking up sharp knives in the kitchen, trying to negotiate a dangerous set of stairs or deciding they want to go for a drive, then it is clearly sensible to try to wake them.

If sleepwalking becomes a more serious problem, it can

be treated with hypnotherapy or even sedatives. A simpler solution is to have more sleep, as a link to a lack of sleep has been suggested. Better-quality sleep is also recommended, by avoiding alcohol and fixing regular bedtimes, so that sleep is less likely to be fragmented or disturbed. However, sleepwalking might also be something to which people are genetically predisposed, as sleepwalking parents are more likely to have sleepwalking children.

Sleepwalking is usually a rather low-key ramble, perhaps involving only pouring a bowl of cereal at three in the morning, but it can also take on more elaborate dimensions. A cook in Scotland had to get help from his doctor when he was getting up each night to make an omelette and chips. More ambitiously, a fifteen-year-old girl in south London had to be rescued after she climbed a 130-foot crane in her sleep, and a hotel survey reported that walking naked to the reception desk was a popular activity for sleepwalking guests.

The bizarreness of sleepwalking has always made it a talking point. In November 1787, the front page of *The Times* featured a stranger-than-fiction story, under the headline 'Remarkable Account of a Sleep Walker', of how sleepwalkers were able to compose music, write sermons and swim without any awareness of their actions. It also mentioned the certain cure that one sleepwalker had employed: 'He threw himself out of a two pair of stairs windows, broke his arm, and never afterwards felt the least return of the disorder.'

See also Snoring, p. 200

Christopher Columbus brought back tobacco, turkeys and the hammock from the new world.

Sleeping to remember

THERE HAS BEEN an increasing amount of evidence linking sleep to memory, suggesting that sleep helps to consolidate memory and learning. The reverse of this, that depriving people of sleep limits their ability to remember, also seems to be true.

Research in recent years has been trying to establish how this might affect learning. Because it's not only that tired people are less successful at remembering, but that the process of sleeping seems to be part of a physical process of turning an experience or a skill into something that we retain as a memory.

Experiments have shown that when people are taught a skill or asked to learn a sequence and then are allowed to sleep before they are tested, this sleep seems to improve their ability to remember. The intervention of sleep increases recall, rather than diminishing it. Sleep seems to provide the glue that helps memories to stick.

Neuroscientists in Switzerland have produced studies showing that a good night's sleep strengthens the learning process, making a discernible difference to the effectiveness of the brain's ability to retain information. In an experiment a group of sound sleepers did much better in memory tests than a control group who had had their sleep interrupted. A study from the University of Dusseldorf, published in 2008, claimed that even a sleep of as little as six minutes can improve the memory. In this test, students were asked to memorise sets of words, and those students who were then allowed to have a short nap proved to be more successful in subsequent memory tests.

The link between sleep and memory is possibly why children need so much sleep. Children have an enormous amount of new information to take in, they are constantly learning new skills, and it's been suggested that this backlog of new experiences is processed in the long sleeping hours. Memory problems faced by older people are also being studied in the context of the relationship between sleep and memory. Researchers have been trying to find out why some older people find it increasingly hard to learn and to retain new memories.

zzzzzzzzzzzzzzzzzz

A study from the University of Dusseldorf, published in 2008, claimed that even a sleep of as little as six minutes can improve the memory.

Experiments carried out on the brains of young and old rats found that the lively young rats were consolidating their memories of the pattern of a maze during sleep. Researchers measured the neural patterns of the rats as they negotiated the maze – and then once again when they slept. The young rats seemed to reproduce similar neural patterns to the ones they displayed when they were in the maze, suggesting that they were 'playing back' this information as they slept, embedding it as a memory that would help them the next morning when they were hunting for food.

The poor older rats seemed to struggle with this task, unable to learn the pattern and unable to absorb it as a memory. This suggested that if the old rats kept getting lost in the maze the next day it was because there was something in the quality of their sleep that prevented the processing of memories. Substitute the maze for the aisles of a supermarket, and I think we get the picture.

For anyone studying for exams or whose children are trying to learn something for school, these are serious findings. When young people cut corners on sleep, they are reducing their capacity to learn.

See also How much did Edwardian children sleep?, p. 107

Sleep and death

LOOK INTO THE FACE of sleep and look into the face of death and, according to Greek mythology, you are looking into the faces of twin brothers. The god of sleep, Hypnos, is the twin brother of the god of death, Thanatos. These brothers are sons of Nyx, goddess of night. It's a family tree heavy with symbolism: death and sleep, the children of the night, looking so similar but with such a separate identity. Sleep is the bringer of peace and healing, his twin brother the destroyer of life.

It's not difficult to see the family resemblance. Sleep and death both involve a stepping away from consciousness, and sleep has long been seen as a temporary rehearsal for the longer departure. There is something about both states that is outside our control, with sleep the benevolent visitor and death the dangerous stranger. It's a kind of mythological good-cop / bad-cop relationship.

The connection between sleep and death has been examined often by writers, as though sleep were a living metaphor for understanding the meaning of death. Shakespeare returns to this theme again and again. Shakespeare's words 'to sleep: perchance to dream' in the famous 'to be or not to be' soliloquy in *Hamlet* have been quoted so often that their meaning has almost been forgotten. But Hamlet weaves his thoughts of living and dying around the 'sleep of death', uncertain whether death might offer the respite that sleep brings.

Large families, early deaths, sudden and incurable illnesses, violence, lack of hygiene and poor medical skills all meant that death in the Elizabethan era was part of life in a way that would seem very unfamiliar to us. So poets looked into the face of sleep to make sense of the ever-present stalking presence of death.

John Donne, in his poem 'Death Be Not Proud', describes death as a short sleep before eternal life. Death should not daunt us, because we have seen its image in sleep, and beyond this final sleep, redemption is waiting.

This idea that death can be described using the language of sleep is carved into headstones in graveyards in phrases such as 'Rest in peace', and appears in obituary columns, with their coy references to someone having 'fallen asleep'. More grandly, the sculptured tombs of the powerful have statues of the interred depicted as though lying asleep in their beds. If you look at the great royal burial places in Westminster Abbey, the Tudor tombs resemble stone four-poster beds. This is a physical impression of the idea that even though death takes away life, the spirit is not extinguished, but, like the sleeper, waits to be re-awakened. The word 'cemetery' embodies this meaning. Although it has come to mean a place for burying the dead, the Greek origin of the word, 'koimeterion', meant sleeping place or dormitory. The cemetery became the place where the dead were sleeping.

There is something irresistibly instinctive behind the Greek myth that sleep and death are so closely connected. Sleep is 'death's second self' says Shakespeare, and it is common for people who have been bereaved to meet their dead loved ones in a dream. These meetings in sleep can feel painfully and reassuringly real.

It's all there in the Greek myth. The gods of sleep and death are imagined spending their endless days together in a darkened cave, beside a riverbank filled with drowsy poppies and sleep-inducing herbs. Their mother, the night, is there to protect them. Beside them runs a river of oblivion and forgetfulness.

See also Dreamland, p. 217

Acknowledgements

Thanks must be paid to my wife Estelle and daughters Anna, Maeve and Josephine for helping me to realise how much I enjoy my sleep. And to my sister Maura for scaring me with her Hansel and Gretel bedtime book. Thanks to the creativity and support of Tim Bates at Pollinger, the picture research and design of Melanie Haselden and Peter Ward, and the improving influences of Caroline Pretty, Nicola Taplin, Stephen Dumughn and Trevor Dolby at Preface.

Picture Credits

All photographs and illustrations courtesy
Mary Evans Picture Library: